Special Needs in Technology Education

A Resource Guide for Teachers

Martin R. Kimeldorf

With contributions from Jean Edwards, Sheldon Maron, and Peter Wigmore.

Davis Publications, Inc.
Worcester, MA 01608

DEDICATION

Nature always sides with the hidden flaw———Murphy's Law

This book is dedicated to my father, Don, and my mother, Fay, who always attended to my special needs and hidden flaws.

Davis Publications, Inc.
Worcester, Massachusetts U.S.A.

Printed in the United States of America
Library of Congress Catalog Card Number: 83-71907
ISBN: 0-87192-148-0

10 9 8 7 6 5 4 3 2 1

CONTENTS

ACKNOWLEDGEMENTS

Because this book deals with an interdisciplinary topic, it evolved through the cooperative efforts of experienced and authoritative teachers in many specialties. Jean Edwards contributed materials on normalization principles, public laws, and individualized education plan's for chapters 1 and 2. Sheldon Maron contributed materials on sensory impairments, and Pete Wigmore contributed material on physical impairments for chapter 3.

Special thanks are due Professor Paul DeVore and the publishers, both of whom recognized the need for such a text and encouraged the author to proceed. Special thanks, too, go to my wife, Judy, whose editing was invaluable.

INTRODUCTION

The purpose of this book is to assist teachers in their efforts to bring into the technological fold one group of students who have been largely excluded: the handicapped. The importance of such a move is understood by the current descriptions of our society as "postindustrial" and "technologically advanced." These terms imply that education must include an involvement with technology. Citizens interact with technology from a variety of different standpoints: production, consumption, recreation, politics. Many handicapped citizens have been excluded from these roles and processes. This exclusion is often based on prejudice, stigma, environmental barriers, and lack of adequate preparation rather than on a person's abilities or assumed disabilities. The net result for the citizen population labeled *handicapped* is unemployment or underemployment and custodial or institutional care. It is to be hoped that the future will bring more alternatives for integrating the handicapped into society.

One of the pathways to integration runs through the middle of the local school system. If curricula and teaching begin to reflect the educational needs of special learners, the dividends of a productive citizenry will far outweigh the initial investment. The schools need to develop "life-centered" educational models. Technology education must certainly play a contributing role, if not a major one, in this sort of educational program.

A major point stressed in this text is that a continuum of educational services, e.g., in-service, interdisciplinary efforts, adaptation of instruction or laboratory settings, remedial or tutorial services, and vocational evaluations, needs to be developed if "mainstreaming" or the goal of a "least restrictive environment" is ever to be fully realized. In this regard, the text attempts to address two main themes:

1. The blending of two fields: cooperative and creative solutions that combine the expertise of special and technological education.
2. The adaptation of instructional methods: practical modifications of proven instructional techniques that will benefit all learners.

Illustrations, photographs, listings, tables, forms, and sequences are included to demonstrate that the blending of two fields and the modification of technique is a pragmatic, attainable goal. Many classroom instructors, project supervisors, and professionals nationwide contributed their advice and materials. The author hopes that the text can be utilized by a variety of professionals (special and regular teachers from any hands-on or experience based fields—art, home economics, recreation: instructors in courses related to technology; professionals in rehabilitation and training programs), as well as parents and students.

The major thrust of this text is to encourage teachers and professionals to try new approaches and techniques related to the instruction of exceptional students. The main barrier is limiting attitudes. As a former industrial arts instructor pointed out, "If you are not willing to see that blind persons are more like you and me than different; if you persist with stereotypes that because someone is blind he will stick his hand in the table saw, then you'll never get off first base." This instructor, Chuck Young, received his formal training in industrial arts. He began working in a wood shop program for the Oregon Commission for the Blind. He is the first to point out that it was his acceptance of a challenge, his imagination—his attitude—that enabled him to succeed. Today Chuck Young is the Administrator for the Oregon Commission for the Blind.

At the close of the twentieth century, a century dominated by our technological creations, we might reflect upon the words of a painter (author's note: identity unknown). This artist applied the paints to his canvas at the dawn of the industrial/technical era nearly one hundred years ago. Looking upon the budding landscape of steel, concrete, and glass, this nineteenth century painter wrote, "Either we all arrive, or none of us will." Whether or not we have created a meaningful technological society will be found in the measure of all citizens' accessibility to this technological landscape. Every person left behind in the shadow of deprivation or degradation becomes a statistic pointing to our failure to coexist with the technology we created, our failure to bridle and harness its dynamism. Every person we bring into the mainstream of our community becomes a hymn to the creativity and wisdom of our labors.

Martin Kimeldorf
1982

Chapter One

BLENDING TWO DISCIPLINES

What role can the instructor in an industrial education setting play in preparing students with special needs to participate in the community to their full potential? More specifically, what sort of support services and instructional techniques can this instructor utilize to make the education of a handicapped learner appropriate? These are the key questions addressed in this text. The answer must be found in the cooperative efforts of two distinct fields: special education and technology education. This interdisciplinary approach is both exciting and challenging. The excitement comes from sharing techniques. On the one hand, the industrial arts teacher will ultimately broaden his or her repertoire of teaching strategies that can be applied to all learners. Likewise, the specialist in special education can enrich his or her curriculum with the knowledge of technology and related technical skills.

PARALLELS BETWEEN TECHNOLOGY EDUCATION AND SPECIAL EDUCATION

Blending the efforts of two distinct professions involves overcoming several related problems: differing vocabularies, methods, content, and teacher preparation. To bridge these disparate elements, this chap-

Much of the material in this chapter was contributed by Jean Edwards.

ter attempts to identify philosophical and methodological topics that parallel each other. For example, the movement in technology education towards interdisciplinary studies is paralleled in special education by the use of interdisciplinary teaching teams to identify the special needs of students.

Normalization is the rationale for mainstreaming. In this text mainstreaming is broadly interpreted to include the minimal adaptations needed for a student with a mild disability as well as the highly specialized adaptations needed for a student with severe disabilities.

The changes that have taken place in technology and their consequent impacts on special and technology education curricula could lead toward an expanded mission for both disciplines. Furthermore, these enlarged professional self-concepts can become the basis for dovetailing efforts—for blending two fields. The result, it is hoped, will be a continuum of educational opportunities that can meet the needs of a variety of learners.

These possibilities are illustrated by this list of the major new topics being explored in technology education and paralleled with the activities and instructional programs found in special education:

COMPARATIVE CURRICULA

Technology Education

1. Study of energy
2. Study of ecology and its relationship to technology
3. Study of consumer roles in advanced technological society
4. Future studies
   Forecasting and coping with technological changes,
   politics of technological development,
   appropriate technology
5. Interdisciplinary approaches to the study of people, science, and technology

Special Education

1. Conversation at home/work
   Maintenance of biomedical (personal) power supplies
2. Work experience programs
   Landscaping, salvage/recycling, grounds maintenance, horticulture, animal care
3. Daily living curricula
   Use of money, personal finance, home and family life training
4. Life/career roles in the community
   Creating a training and service continuum to support deinstitutionalization and survival in the community
   Biomedical technology and rehabilitation engineering
   (prosthetics, orthotics, sensory aids, biofeedback, etc.)
5. Mainstreaming
   A program for providing the least restrictive and most appropriate education involves an interdisciplinary effort for program design and implementation.

COMPARATIVE EDUCATIONAL ACTIVITIES

Educational Domain	Technology Education Activities	Special Education Activities
Cognitive	***Cognitive***	***Cognitive***
1. Learning how materials and processes are grouped, organized, categorized, and labeled.	1. Identifying materials and sequences in fabrication, assembly, finishing, etc.	1. Reading skills: decoding and comprehending material.
2. Academic skills	2. Applied use of math/measuring, reading, and writing	2. Remedial instruction in math, reading, and writing
Affective	***Affective***	***Affective***
1. Self-expression	1. Projects in design and fabrication	1. Written language instruction, speech therapy
2. Self-concept	2. Projects: completion, display, mastery of useful skills	2. Grades, reward systems (special awards, recognition)
3. Coping with stress	3. Hands-on outlet/therapy	3. Behavioral programs, counselling
4. Work habits	4. Learning safety and correct work habits	4. Work adjustment/work experience programs
5. Independence and socialization	5. Home repair skills, career education, opportunities to make friends (i.e., participate in a buddy system, club, mass production project requiring group efforts, personnel system)	5. Daily living training, career education, social/sexual education, counseling, special arts festivals, special Olympics
Psychomotor	***Psychomotor***	***Psychomotor***
1. Coordination and strength	1. Use of tools and materials	1. Adapted physical education, recreational therapy, physical and occupational therapy
2. Perceptual abilities	2. Use of plan sheets, blueprints, models; use of design principles	2. Remedial or specialized instruction
Specific Disciplines	***Specific Disciplines***	***Specific Disciplines***
1. Communication systems	1. Electronics, graphics/drafting, video/film, laser	1. Personal communications: communication systems for language, hearing and seeing impaired; bio-engineering (using transducers to control machines); and use of prosthetics
2. Transportation	2. Power mechanics, fluid control, energy, mass transit, rocketry	2. Access to buildings and transportation, adapted personalized transportation (wheelchairs, adapted vehicles)
3. Production	3. Woods, metals, electronics, computerized processes, impact on labor and leisure	3. Vocational habilitation and rehabilitation (work experience and training)
4. Avocations	4. Hobby skills, crafts, home and auto repairs	4. Adapted physical education, Special Olympics, special arts festivals, instruction on using leisure time, camping/outdoor programs

The exceptional student can benefit and grow towards greater independence by participating in *both* curricula.

A more specific breakdown demonstrating how the two curricula can dovetail or complement one another is presented on page 3. Providing the learner with generalized repetition, an additional opportunity to apply and generalize skills, can mean the difference between a relevant and appropriate education and one that only pays lip service to the needs of handicapped students. Exceptional students, like regular students, cannot consider their education complete if it does not at some point include a study or experience in technology.

The sum total of these endeavors can mean a more promising future for handicapped citizens. Many demonstrations and on-going projects in schools and communities show unequivocally that handicapped citizens can contribute to their communities when they receive appropriate educations. Each day the foggy mist of stereotypes, myths, stigmas, and half-truths is being lifted as citizens with disabilities begin to participate to their full potential in our communities.

Our society must adopt a more humane outlook. We must accept and nurture all of our citizens. Ultimately, this will happen, not because of the technology we possess, but because of the democratic ideals we practice. The most important educational method of all is our attitude.

STIGMA

Key to the blending of these two fields is the instructor—and the instructor's attitudes toward handicapped persons. Technology education is currently hampered from adequately meeting the needs of handicapped students via mainstreaming not only because of "professional turfdom" but also because of the stigma attached to special needs students. These attitudes are sometimes fostered at the administrative level by those who resist integration programs, citing limited fiscal resources and insufficient numbers of specialists to meet the needs of this special population. Such attitudes are often deep-rooted.

Consider the prevalent attitudes toward handicapped persons. One person might believe that handicapped people are strange or frightening. Another might believe that handicapped persons are subhuman, menaces, objects of pity, or eternal children. Yet another might believe they are no different from other people. Positive and complimentary perceptions are accompanied by liking and acceptance; negative beliefs by dislike, fear, and avoidance.

In the public school, stigma interferes with the educational process in many ways. First, stigma has resulted in the segregation of special needs students from the mainstream of public education. Stigma has limited the special needs student from partaking of the offerings of social, vocational, and recreational options afforded regular students. Stigma has resulted in special needs students' peers regarding them as odd, strange, and objects to be avoided.

A study reported in the *Wall Street Journal* indicated that not only did peers carry stigma that resulted in handicapped students being isolated and rejected, but that regular class teachers also harbored many stigmatizing fears. Teachers reported that they feared special needs students because they were odd and different. Teachers feared integration of special needs students in the regular school program because they feared being unable to successfully teach these students. Further, they feared disruption, peer pressure, ridicule, and their own devaluation because of not being able to successfully teach such students. The study also reported that public education alone is not effective in changing teacher attitudes. Data obtained from this study indicated that increased information resulted in about 50 percent of the teachers making a positive change in attitude and increased acceptance of handicapped persons. More effective was direct contact. Those teachers and students who interacted with special needs students in school, in the neighborhood, at work, and in leisure activities showed a greater acceptance of these special needs persons. Many reported discovering that these "handicapped people are more alike than different" from nonhandicapped people. Teachers with direct contact acknowledged the special needs of these students and were able to see the students as more like their regular students than different. Attitude modification is difficult and sometimes unsuccessful. Simply giving people information is not enough. Research indicates that active participation, as in role playing, volunteer efforts, or guided practicum experiences at the university or in-service training level for teachers, is more effective in changing attitudes than is passive exposure to persuasive communications.

Stigma leads to isolation of handicapped students, and isolation deprives them of important developmental experiences that ultimately could open the door for successful vocational participation. Many handicapped students may have experienced a delay in the acquisition of basic technology education (e.g., a lack of familiarity with tools because it was feared they would injure themselves). In this way we find that stigma can lead to a self-fulfilling prophecy about the skills of handicapped persons. So, hand-in-hand, regular educators and special educators must work to remove the stigma that creates the isolation. This process is imperative to the normalization/mainstreaming goal of a life-centered technology program.

NORMALIZATION

In the United States, the blending of special education and technology education began some ten years ago with a movement to provide more normal experiences and life-styles for people with special needs. This movement has cut across educational, social, vocational, and residential services for special needs students. The movement began in Scandinavia where they adopted a normalization principle in the 1960s and firmly embodied their ideology in legislative action to change educational, residential, and vocational services provided for their handicapped and special needs populations.

Normalization is simply:

> ...making available to the handicapped, patterns and conditions of everyday life which are as close as possible to the norms and patterns of the mainstream of society (Nirje, p. 181).
>
> ...utilization of means which are as culturally normative as possible in order to establish and/or maintain personal behaviors and characteristics which are as culturally normative as possible. (Wolfensberger, 1972, p. 28).

More popularly known in the United States as "mainstreaming," normalization has swept across the educational and social service systems that serve special needs persons in our country and has resulted in the integration of thousands of formerly segregated students in regular educational and technology programs. Normalization and mainstreaming found their expression in Public Law 94–142, the right to education law for all handicapped and special needs children. Integration is one of the most significant attributes of normalization, and while many professionals readily endorse the principle, they engage in practices that oppose mainstreaming.

Public Law 94–142 has been hailed as the "Bill of Rights for the Handicapped." The law affirms "a free and appropriate public education and related services designed to meet their unique needs" (section 601(c)). Further, the law provides that this education should be received in the least restrictive environment possible. Special educators, vocational/technology instructors, classroom teachers, and other school personnel are presently redefining their respective roles and responsibilities as they relate to services for the special needs population. Federal regulations for P.L. 94–142 promulgate that:

> Each public agency shall take steps to insure that its handicapped children have available to them the variety of educational programs and services available to non-handicapped children in the area served by the agency including . . . industrial arts, consumer and homemaking education and vocational education. (Section 102(a), 305)

Public Law 94-142 gave normalization the teeth it needed. The normalization principle requires—in addition to all the typical practices, structures, and interpretations—that each student with special needs receive services that provide the special programming required for maximum development. In order to insure the benefits of normalization, the public law was essential. Also, without the requirement of an individualized educational plan in the public schools, normalization could be little more than a dream.

In the past decade normalization has also found expression in the surge of new programs for special needs students. The most relevant and significant was in the development of vocational services for special education programs in the 1960s. The emergence of work-experience programs, job-training skills being taught in schools, and community work-training stations in the community are all results of the growing awareness of the best conditions in which special needs students could learn. Research and training projects validated the concept that the normalized community-based training environment increased the retention of the vocational skills taught.

Avocational learning experiences emerged in importance as community experiences took on greater importance in the public schools. As curriculum design committees met to develop career and vocational curricula, educators began to see the interdependence of recreation and leisure skill development, social-sexual training, the acquisition of daily living skills, and technology education curriculum. Technology education and normalization can help define the specific content of a life-centered curriculum, thus helping to implement P.L. 94-142. At the same time normalization can involve parents, students, regular educators, and special educators in a partnership with unlimited potential.

The Vocational Education Act of 1963 and its amendments in 1968 mandated that 10 percent of the federal funds allocated under part B of the act be designated to provide vocational education to the handicapped. Other legislation and federal allocations brought millions of federal dollars to states and school districts to improve the quality of special and vocational education. Many gains were made for special needs students through these legislative actions. At the same time, however, Oregon researchers Espeseth and Hammerlynck in 1969 documented the problems incurred when the lack of communication between vocational educators, special educators, and rehabilitation personnel occurs. Special needs students lose out in the midst of our dilemma over terms and professional territoriality, turfdom, and jealousy.

Finally, the regulations of Section 504 of the Rehabilitation Act of 1973 (P. L. 93-516) set forth requirements for nondiscrimination on the basis of handicap in preschool, elementary, secondary, and adult education programs and activities. One of the principal objectives of section 504 is termination of discriminatory or inaccessible programs in public schools. A crucial distinction must be made here: Section 504 does not order school facilities to be totally free of architectural barriers. Section 504 does order program accessibility. Therefore, since June 1, 1977:

(1) All new facilities constructed must be constructed so as to be readily accessible and usable from a program standpoint by all handicapped students.

(2) Although the total school need not be totally physically accessible, there must be an assurance that programs conducted in those facilities are made accessible.

(3) Structural changes should be made "expeditiously" to make programs accessible (subpart C of section 504 regulations).

Section 504 furthered the causes of normalization by requiring this accessibility. No longer could handicapped students be denied technology programs access because buildings were not accessible.

School districts, state departments of education, teacher training institutions, legislators, and others are now attempting to define and direct these mandated services for the special needs population.

TECHNOLOGY EDUCATION: PATH TOWARD NORMALIZATION

Special needs students can benefit from a technology course in a variety of ways. Here are some of the potential advantages.

1. Because it is a realistic environment, the technology laboratory can yield useful information about a student's abilities in specific terms. For example, the student's ability to be responsible with materials, to get along with others in a work or hands-on setting, to use tools, to measure, and to operate machinery all can be observed. This stands in contrast to the information that certain vocational evaluation systems yield. For example, some of these systems only describe the student's abilities in such global terms as "motor coordination, "work rhythms," and "work quality (errors)."
2. Because it is a realistic environment, and replicates the "outside" world, the technology lab is highly motivating. This is evident with nonhandicapped students as well. For the special education student who often is taught in a contained classroom or curriculum utilizing role playing, simulation, and abstract puzzles, the technology lab offers a refreshing involvement with real materials and processes. The lab setting often validates whatever else the student is learning.
3. Learning deficits need not be a source of stigma or lowered self-concept. Technology labs have often served students with a variety of abilities and talents. Often, nonreaders and gifted students may be found in the same or similar programs, especially in exploratory or individualized programs. The technology instructor is adept at accommodating students' academic skills in lab works.
4. Remedial instruction, which often consists of fragmented and isolated skills taught in structured formats, can be related to technology processes. Thus their purpose and mission can be enhanced.
5. Modifications made to accommodate special learners need not be extensive. Special education entry-level courses can be easily adapted from beginning exploratory classes or mass production classes. Additionally, these classes can serve a vari-

ety of students not in special education. Likewise, materials developed to accommodate instruction for slower learners are often utilized by all class members (thus the materials need not be associated with a stigma).

6. The technology lab affords an opportunity to develop one's interpersonal and work skills. The buddy system, personnel systems, and sharing modes prevail in most labs. Additionally, the student must learn and demonstrate responsible attitudes toward the use of materials, equipment, and time. This often stands in contrast to traditionally academic resource or remedial special education classes that are highly structured as regards behavior and the use of time.
7. The technology class can become a renewed source of self-esteem. Inherent in an environment of risk is the dignity of working as others do, without being overprotected. Similarly, the useful skills learned or projects made can be shared with others, thus demonstrating one's abilities, not disabilities.
8. The technology curriculum can augment physical and occupational therapy programs. For example, a student who suffered serious brain damage in an accident pursued a program in the business field that included use of the hand tools used in offices. To relearn cutting with scissors and to build strength, the student's program consisted of starting with tracing outlines, cutting out these outlines with scissors, and eventually using sheet metal and tin snips to build up reserve strength.

TECHNICAL LITERACY

The study of technology has undergone an expansion concurrent with the rapid proliferation of man-made hardware. This study of technology (and its impact on society and persons) has given the very concept of technology new meaning. The general reader may have come across such names as Marshall McLuhan, Alvin Toffler, and Harry Braverman. Harry Braverman's in-depth study of the impacts of technology on the work place and worker represent a germinal landmark work. He is the author of *Labor and Monopoly Capital.*

The expansion of our technologically based culture has had a similar impact on special education. Professionals in this field have expanded their definition of their field of inquiry and service at about the same historical moment. Special educators have borrowed from the behavioral and hardware technologies. For example, the principles of behaviorism have been successfully interpreted and applied as a teaching technology. In addition, the use of high technology electronics has been applied to produce low-vision aids, reading machines, and sound sensing and amplifying devices for seeing and hearing impaired people. The use of drugs, though controversial, also has its roots in the advances in biochemistry and related drug therapies.

We live in an era of continual metamorphosis. The winds of change have stripped away the traditional image of a shop teacher as an organizer of nuts and bolts and producer of venerated Christmas projects. The technology instructor has emerged as an interpreter of technology in an advanced industrial society.

Delmar Olson, an early proponent of technological literacy, stated "until each of its teachers becomes a philosopher, IA can amount to little more than busy work." In laying the groundwork for an expanded vision of technology, he further wrote about the expanded mission for industrial arts as including:

- Technical literacy
- Consumership
- Recreational expression
- Cultural efficiency
- Personal self-realization and release (1974)

The implication here is that industrial arts education is essential for all learners because it can serve the total needs: vocational, avocational, social, and scientific. Olson elaborated on this theme thus:

> Industrial arts can round out the academic educative treatment of the student in the sense that developed hands are essential to a complete education and undeveloped hands means but a partial education (1973).
>
> Man as an individual is by nature a creator and maker. He uses his hands and his imagination as a means of self-expression. Once found in work, he now gets it in his leisure (1974).

Technical literacy implies a broadening of the commonly offered courses. For example, transportation clusters would replace mechanical power clusters. With the transportation cluster the student not only studies specific mechanical functions and technical information, but takes into view various modes of transportation (e.g., rocketry, mass transit, motorized wheelchairs) and surveys their relative impact on ecology, society, and urban design.

The very notion of what constitutes literacy in a technological age is now in question. The concept of a literate person as one who can read (decode words, comprehend sentences) and knows civics (historical facts and sequence) is perhaps too limited today. Literacy includes new language skills (use of photographic and computer techniques, blueprints, flow charts, etc.) and an expanded awareness of the social impacts of technology (e.g., technological forecasting, future studies, political and social implications of technology, influence on contemporary aesthetics and design).

Perhaps literacy or the study of technology will be redefined in the 1980s. In Cabell County (West Virginia) the school district is attempting to field test a unified arts program. In a school in New York, educators have coupled science and industrial arts areas related to graphic arts photography by photographing life under a microscope. Interdisciplinary efforts can utilize more than two fields. For example, theater can be coupled with special education and referenced to career education. (See *Don't Get Fired* by Durlynn Anema). In another example, a combination of the visual arts with the graphic arts offers a comparison between orthographic projections and Cubism. (See *Industrial Arts-Integrating Man/Society/Technology* by Martin Kimeldorf).

Paralleling this trend to broaden the definition and scope of the technology education field is the movement toward life-centered curricula in special education. In the past, special education was primarily an isolated program of remediation in the basic academic subjects (math, reading, writing) coupled with the standard therapies (physical, occupational, speech). Today we find a plethora of emerging therapies—music, art, recreation, dance, adapted Physical education. The broader mission of special education is, according to Donn E. Brolin, an expanded curriculum, a total career program that would include "one's role not only as a producer [of work], but also as a learner, consumer, citizen, family member, and social-political being" (Brolin, 1973).

A life-centered curriculum is defined by Robert N. Freeman as "a series of planned experiences which prepares persons to live well in their environments . . . and should be considered different from 'survival skills' which are meant only to train persons to 'get by' on a day-to-day basis" (Freeman 1980).

A life-centered curriculum is a series of planned educational activities that will prepare the student to operate appropriately, independently, and competently in school, the home, the community, and in future work. The skills taught in technology education will enhance this life-centered curriculum.

In the life-centered curriculum, teachers are required to look at the individual student's lifetime needs and develop a program within his or her area of competence. It should not be an attempt to fashion a program that fits the needs of a high school diploma unless the needs relate to the realities of the community and are relevant to the life needs of the individual student. The concept of a life-centered curriculum is not new in education, but its implementation has been little more than talk.

The life-centered curriculum is needed for all students, but it is essential for special needs students. Each person has the right to be all he or she can be. A life-centered curriculum opens the door to success for all students by providing flexibility in educational experiences. These types of curricula typically include training in community travel and orientation, daily living skills, home management and consumer skills, social-sexual education, recreation, and vocational skills. The curriculum provides special classes as well as support for related regular class offerings. The key concept here is what Freeman calls "a series of *planned* experiences." Without a concerted and comprehensive plan, the attempt to implement a life-centered curriculum is similar to the attempt to cross a mine field without a map.

To create this life-centered curriculum, special educators and regular educators need to blend their efforts. Of major importance is a shift from the traditional content-based curriculum to one that relates directly to the outside world and focuses on each student's unique ways of learning and being motivated. The development of practical skills rather than knowledge and information must be emphasized. Furthermore, the partnership will have to extend beyond the school and into the community.

Life Centered Technology Education Model

In this section an interdisciplinary model is proposed that attempts to delineate the role that technology education can play within a larger life-centered school curriculum. This Life-Centered Technology Education Model is considered timely in light of the changes emerging within the respective disciplines, and with regard to the mandate for a normalizing educational experience.

A special education student can benefit from placement in a regular class when a continuum of educational services prepares him or her for placement and carry through after the regular course has ended. Some students will need remedial, specialized or alternative placement options as described in chapter 2. The model shown on page 13 was designed to serve the full range of disabled students, from the mildly to severely handicapped. It attempts to portray two main themes of an educational continuum:

1. The placement and growth toward the least restrictive environment should progress in an orderly fashion from specialized classes (as needed) toward regular classes terminating in services leading to independent living in the community.
2. The progress within the technology curricula should be accompanied by related curricula. Preparation directed toward community-referenced goals will be most efficient when the overall student program is coordinated.

Three instructional stages are portrayed: special education, regular education, and postsecondary education. While each appears to have a separate domain, this division is made only for convenience. These programs must be linked in terms of transitional services. Specific examples of these transitional services are described in chapter 5.

The model is a menu listing many options from which educators and staff can choose from. The options include the conventional as well as innovative course and curriculum possibilities. The model embodies the concept of a learner progressing from the most specialized environment (Special Education-Entry Level) towards the least restrictive (Technology Curricula) and culminating in the community mainstream. The concept of a

LIFE-CENTERED TECHNOLOGY EDUCATION MODEL

Special Education	Regular Education		Post-secondary community and Educational Services
Technology Curricula	***Technology Curricula***		***Technology Curricula***
Entry Level	*Exploratory Level*	*Cluster Level*	*Vocational Training*
Preentry Class or Vocational Evaluation Pre-vocational Training • Preauto/Metals/Woods/etc. • General Technology Lab	Industrial Arts • Beginning Classes Applied Skills • Home Repair • Hobbies/Crafts • Mass Production	Vocational Clusters • Unit Shop: Auto, Woods, Metals, etc. • Occupational: Power Mechanics, Graphics, Construction, etc.	Sheltered Workshop/Training Vocational Rehabilitation Services Specialized Training Programs and Placement Services with Community Emphasis
Applied Academics • Technology Math/English/etc. • Career/Social Studies		Technology Clusters • Transportation, Communication, Production/Manufacturing, etc.	*Adult Education*
Interdisciplinary Classes • Team teaching with instructors from special and regular education Work Experience (Supervised) • Classroom/Campus Projects • Classroom/Campus Work Stations • Special Projects • Class Co-Op Projects	Interdisciplinary • Technology and Science • Technology and Art • Technology and History • Technology and Society	Individualized • General Lab: individualized goals • Specialized Classes: landscaping, salvage, custodial, etc.	Specialized Life-Centered Educational Programs Specialized College Settings, e.g. National Technical Institute for the Deaf Modified Curricula in Community Colleges Regular College Placement with Support Services, e.g. Handicapped Student Services, Affirmative Action
Related Curricula	***Related Curricula***		***Related Curricula***
Mobility • Personal Mobility Skills • Community Orientation Personal/Social Adjustment • Personal Appearance/Hygiene • Social/Sexual Education • Daily Living Skills • Career Education • Recreation (Adapted PE/Therapy, Special Olympics, Art Festivals) • Orientation to Community Services and Laws.	Mobility Driver's Education Geography Personal/Social Adjustment Home Economics Family Life, Biology Bachelor skills, Personal finance Career Guidance Physical Education, Clubs, Extracurricular Sports, Art Classes, Shows Social Studies		Mobility Mass Transit and Specialized Transportation Services Personal/Social Adjustment Group Homes, Supervised Residences Vocational Training, Adult Education Parks and Recreation Department Service Organizations and Agencies

continuum of placement options is critical. To simply place a student in a "regular class" is not mainstreaming. This haphazard approach often retards progress for the student and frustrates the teacher. A placement based on the ability of the student is the goal of "least restrictive educational placement" or mainstreaming.

Schools that want to offer relevant educational opportunities for speial needs learners must strive to provide a continuum of instructional offerings from entry level to exploratory level through cluster and postsecondary transitions. Whether the content is conventional or innovative will depend on the staff expertise and attitude as well as existing resources. Entry level courses have been variously called pre-entry, prevocational, readiness, applied, and basic courses. Such a course offers the student the chance to explore his or her interests and aptitudes. It can augment a vocational evaluation or supplant one entirely if data can be routinely collected on student performance and summarized for placement and training purposes. This course can also be an opportunity to master entry-level skills regarding tool names and uses, identification of materials, practice in processes, and acquisition of necessary behaviors (e.g., personnel systems, safety, work habits, interpersonal skills). It can offer the student a running start in the regular program, build up confidence, and avert inappropriate class placements.

The entry-level program supplements the regular curricular offerings. As such, it often comes under the jurisdiction of special education in terms of promoting, design, and implementation. However, many schools hire a special educator to provide this form of instruction within a special resource room or within the existing lab facilities. Other schools create a special class, which is team taught. One such program is found in the Madison (Wisconsin) Metropolitan School District. This program has an entry level program called The Basic Class. It provides instruction in entry-level skills in five areas: home economics, art, business education, industrial arts, and agriculture.

The next logical movement is placement within the regular curriculum. This has been divided into exploratory and cluster levels. The exploratory level is an important link to future placements because it is the first instance where the exceptional student usually experiences the rigors, risks, and challenges of the regular courses. It is a pivot point that can neatly fit into either the specialized or regular program. The author is of the opinion that this should be an integrated class that can serve a variety of students with different special needs: nonvocational students interested in technology, beginning students, and handicapped students. Examples of this sort of class could be the conventional mass production, home maintenance, or industrial arts courses, or innovative interdisciplinary courses like theater arts technology, technology and history studies, or technology and science.

The next logical step becomes placement in a cluster-level program. It cannot be overemphasized that success in a regular class is directly dependent on the amount of planning and coordinating prior to placement, and subsequent support. Support can take various forms, from tutoring, adapting curriculum, grading, or instructional techniques (see chapter 5 for a fuller description), as well as support for the regular teacher (e.g., in-service, team planning/teaching). Additional services might include consultation and instructional support from behavioral specialists and occupational and physical therapists, and communication with parents and community service providers. Finally, at the end of the high school career, appropriate community agencies must be contacted and brought into the transition of graduation from school to community services.

There are various ways to staff support services related to regular class placements. To date, many occupational and vocational/technical centers rely on a model in which a special education teacher is assigned the responsibilities of coordination and support on a full-time basis. In the opinion of the author, this sort of full-time staffing commitment needs to be duplicated in the comprehensive high school as well. This type of position might be called a technology resource instructor. The time commitments might be equally divided between participation in the technology lab with student and regular teacher, as well as the special classes in the life-centered related curricula.

One example of such a position can be found at Venango County (Pennsylvania) Vo-Tech. Sheila Feichtner and Thomas O'Brien helped to design a program with a staff that included a vocational resource person. His job was to utilize the total resources of the community, plan job-training programs, provide counseling, provide in-service, and create new programs as needed. A representation and description of this staffing model is found in figure 1-1. In project SERVE at the 916 Area Vo-Tech Institute in Minnesota, students are being served by a supplemental resource instructor (SRI). The responsibilities of the SRI are similar to the vocational resource person and described as:

> . . .the SRI are generally not intrusive and the casual visitor to the SERVE program is likely to assume that the SRI is simply one of the instructors in the occupational area. . . . He also mediates whatever curriculum modification is found to be necessary. This may include the rewriting of a learning package, modification of the work place to accommodate a handicap, altering the educational environment so as to enable progress or to promote a particular growth, or other interventions that are required by the needs of the student. Supplemental instruction itself is furnished by the SRI. This is one reason for the specialization, since the SRI must be knowledgeable in the field of [technical] training (SERVE 1977–78).

In the life-centered technology education model, an important option exists that should not be overlooked: individualized or community clusters. This implies a specialized or individualized placement within an existing cluster (see chapter 2 for a discussion of placement options). It also opens the way for short-term or creative courses. An example of this is the

MODELS FOR MAINSTREAMING

In the following models the needed support services are indicated by the three inner circles and the outer circles. The inner circles represent an instructional team of special and regular education teachers. The outer ring represents community and school resources.

UNIT SHOP MODEL

This is the unit shop or technology lab with a single content area (e.g., building maintenance). Remedial instruction for related academics (e.g., math or measuring) is taught by the special education teacher using materials from the vocational setting.

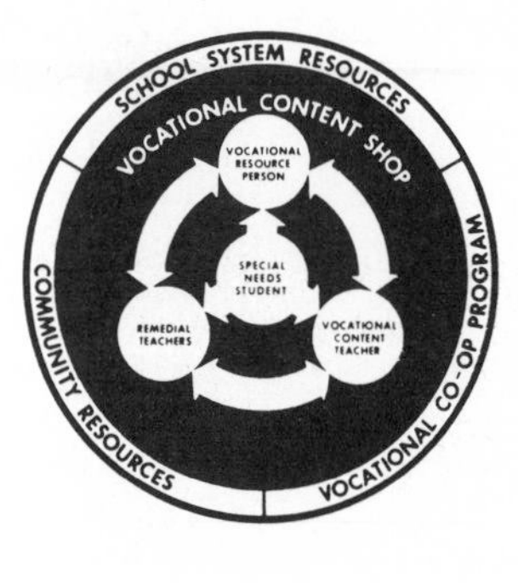

CLUSTER SHOP MODEL

As the technology unit expands in content to a cluster curriculum, the model can remain basically the same. Flexibility is added to student placement options as the technology model expands its curriculum content. For example, a student may enroll in the cluster but specialize in only one area of the cluster (e.g., choose finish work in a building trades cluster).

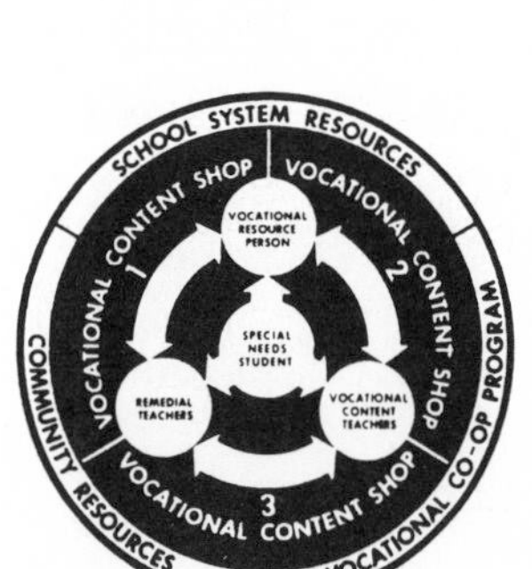

Figure 1–1. (Models courtesy Dr. SheilaGeichtner, Indiana University of Pennsylvania.)

building maintenance course taught at the local community college organized by the Special Services Cooperative in Olympia, Washington. In this example, high school students are excused for up to a half day to attend this intensified instruction in a class not normally offered in the regular curriculum but which is referenced to local community employment options. Courses in the past have included horticulture and landscaping, which can serve vocational as well as avocational pursuits.

Finally, the model touches upon the postsecondary options in the community. For many students (ages 18–21), graduating becomes a new life crisis. This sense of crisis may be heightened for the special needs student and is described in poignant terms by the College for Living, in Denver, Colorado:

> Many handicapped persons are in the middle of the richest learning process of their lives when "graduation" comes and the strengthening and liberating force of special education is abruptly stopped.

The College for Living, described in chapter 5, attempts to fill this gap.

In summary, a continuum of educational services is presented in this model. It begins with a simplified entry-level program geared to the instructional methods of special education but with the content of technology education. This prepares the student for a regular class placement, which may be either partially adapted (e.g., exploratory level) or be enhanced by supportive and supplemental services. The goal is always an integrated and normalized educational setting (e.g., cluster type of class). Finally, a transition must be planned between the secondary program and the postsecondary community program if the entire effort is to bear the fruit of independent living.

Life-Centered Education for the Severely Handicapped Student

Some researchers are of the opinion that secondary programs for the severely handicapped must differ markedly from programs provided for the mildly handicapped. Others have concluded that for many severely handicapped students the major change will be twofold:

1. Utilization of the technology lab as a simulated work setting prior to placement in a community-based training program
2. Utilization of the community as a training environment, whether for technology/vocational training, mobility, living, or leisure education.

In a well-articulated program, the above requirements can be easily incorporated. In reference to the Life-Centered Technology Education Model, a student might spend a majority of his or her prevocational training in an entry-level class, then proceed to specialized placement (e.g., to learn a specific skill) within a regular class that emphasizes personal behavior as well as skill development. Finally, the educational program would terminate in a community setting. Thomas Bellamy and Barbara Wilcox, researching specialized training for severely handicapped individuals, describe the components or program qualities for the severely handicapped as follows:

- Integration
  Services and training programs are delivered in accordance with the principle of normalization, with opportunities for integrated instructional settings.

- Age Appropriate
  Instructional content and training should be determined by age appropriate references and discard notions that the severely handicapped are to be treated as "eternal children."
- Community Referenced
  Objectives for instruction should be based on specific community options for living, working, leisure and training/support services.
- Future Oriented
  The objectives should not be limited to present confinements, but should look towards a future of expanded services.
- Parent Involvement
  Potentially there may be many areas of instructional need which must be delivered within a limited time. Therefore, parents need to be involved as active participants in determining content and priority of the individual educational plan.

The two major implications are that the school services must be comprehensive and that community-referenced instruction should lead to community-based training and student evaluation. Chapter 4 enumerates possible instructional strategies related to serving the severely handicapped student.

EXPLORATIONS

Activities

Visits to:
- Occupational Versatility labs
- Vocational training projects for the handicapped in the community
- Very Special Arts Festival
- Special Olympics

Readings

In addition to the sources cited, the following can broaden one's awareness of technology and special education:

Magazines listed in appendix C, especially:
- *Career Development for Exceptional Individuals*
- *The Exceptional Parent*
- *Industrial Education*
- *Man/Society/Technology*
- *School Shop*
- *NAVESNP Journal*

Texts listed in appendix C, and:

Brolin, Donn E., ed. *Life Centered Career Education: A Competency Based Approach*. Reston, VA: The Council for Exceptional Children, 1978.

Lake, Thomas, P., ed. *Career Education: Exemplary Programs for the Handicapped.* Reston, VA: The Council for Exceptional Children, n. d.

Olson, Delmar, W. *Technol-O-Gee: Industrial Arts Interpreter of Technology for the American School.* (cited in references)

REFERENCES

Anema, Durlynn. *Don't Get Fired! Thirteen Ways to Hold Your Job.* Hayward, CA: Janus Book Publishers, 1978.

Bellamy, Thomas G., and Wilcox, Barbara. "Secondary Education for Severely Handicapped Students: Guidelines for Quality Services." In *Critical Issues in the Education of Autistic Children and Youth.* Edited by B. Wilcox and A. Thompson (in press).

Brolin, Donn E. *Vocational Preparation of Retarded Citizens.* Columbus, Ohio: Charles E. Merrill, 1976.

Cook, Paul F., Dahl, Peter R., and Gale, M. *Vocational Opportunities.* Salt Lake City, Utah: Olympus Publishing, 1978.

Cuneo, Eugene J. "The Interdisciplinary Approach: LA and . . ." *Industrial Education.* 69: 32, May/June 1980.

Freeman, Robert N. "Needed: A Life Skills Curriculum in the Public Schools." *Education Unlimited.* 2:45–46, Feb. 1980.

Kimeldorf, Martin. "Art and Theatre in a Labor-Saving Power Tool Century." In *Take a Card, Any Card.* Manhattan, KS.: Kansas Association for Retarded Citizens, 1980.

Kreps, Alice Roelofs. *Metro College for Living Workshop Packet.* Denver: College for Living, n.d.

Lippman, Leopold. *Attitudes Toward the Handicapped.* Springfield, IL: Charles C. Thomas, 1972

Murray, Michael L. and Murwin, Scott W. "Studying Technology through Unified Arts." *Industrial Education.* 68:8–10, May/June 1979.

Nirje, - in "Normalization Principle and its Human Management Implications" in R. Kugela and Wolf Wolfensberger. *Changing Patterns in Residential Services for the Mentally Retarded.* Washington, D.C. President's Committee on Mental Retardation. 1969.

Olson, Delmar W. "Interpreting a Technological Society: The Function of Industrial Arts." *School Shop.* 33:35–36, March 1974.

Olson, Delmar W. *Technol-o-gee.* Raleigh, N.C.: North Carolina State University, 1973.

President's Committee on Mental Retardation. "Mentally Retarded Persons Go to College." In *Mental Retardation: The Leading Edge—Service Programs That Work.* Washington, D.C.:HEW, Office of Human Development Services, MR78.

SERVE. In Chapter IV *Components of SERVE.* White Bear Lake, MN: 916 Area Vocational Center and Technical Institute, 1977–78.

Starkweather, Kendall N. "The Nature of Industrial Arts in a Post-Industrial Society." *Man/Society/Technology.* 35: 196–200, April 1976.

The Rockwell Power Tool Instructor. Mainstreaming in Action. 23: 3–7, 1978.

Wolfensberger, Wolf. *The Principle of Normalization in Human Services.* Toronto, Ont.: National Institute on Mental Retardation, 1972.

CHAPTER TWO

STUDENT PLACEMENT AND EVALUATION

Commitment to normalization is not only humanistic, it is mandated by law—Public Law 94-142. General background concerning the content and intent of the legal mandate is presented here. Since this law, as all laws, is bound to become revised, and because it is written in fairly technical detail, only an overview is presented. Likewise, only that part of the law that most effects classroom teachers will be discussed. Related to the law is the question of placement options within the regular curriculum as well as the need to provide the least restrictive and most appropriate educational opportunity for the special needs learner as expressed in an individualized education plan. Often this is guided by a vocational evaluation. All of these topics, as well as current problems encountered in mainstreaming, are presented in this chapter.

A BRIEF INTRODUCTION TO PUBLIC LAW 94-142

The passage of the Education for All Handicapped Children's Act of 1975, Public Law 94-142, was a long-awaited event in the struggle for equal educational opportunity for handicapped children and youth, ranging in

Much of the material in this chapter was contributed by Jean Edwards.

age from three to twenty-one. It signaled the beginning of mainstreaming and normalization in the United States on a national basis and was the beginning attempt through legislation and federal funding to reduce stigma and segregation of the handicapped and to foster acceptance and integration of the handicapped students into public education. While the law has many new aspects, it has its roots in federal laws that developed from the late 1950s on and ensures each handicapped student a free and appropriate public education.

The law not only requires "a free and appropriate public education and related services designed to meet their unique needs" (Section 601 of the Act), but it further contains six major principles:

1. Zero rejection of students in public schools
2. Testing and classification as the basis of appropriate school placement
3. Individualized and appropriate education for all
4. Use of the least restrictive appropriate educational setting
5. Procedural due process (i.e., handicapped persons' and parents' rights safeguarded with respect to identification, evaluation, and educational placement whether it be the initiation or change of such placement or the refusal to initiate a change.
6. Parent participation and shared decision-making

Key to the implementation of the law at the secondary school level is the individual education plan (IEP).

Individual Program Plan

The purpose of the individual education plan is to move handicapped students toward more independent and productive living. Because they contain goals, objectives, and timetables for achieving them, IEPs can help determine whether or not schools are providing effective services. The law provides a way of making special education and technology educators accountable and helps teachers to become conscious of achieving the goals with the learner.

According to Public Law 94-142, an IEP must:

- Be written
- Describe the student's present levels of educational performance (evaluation, diagnosis)
- State annual goals
- State short-term instructional objectives
- Describe specific educational services to be provided
- Determine the extent of the student's ability to participate in regular educational programs

- Determine the starting date of the student's program
- Anticipate the duration of services
- Select appropriate objective criteria and evaluation procedures to determine whether instructional objectives are being achieved
- Determine the schedule for evaluating progress, at least annually

The IEP is basically an agreement between (a) The consumer (the handicapped student) and his or her parents or guardians, and (b) the service provider (the school) and the teacher and resource specialists available to provide supportive services.

Types of Information

Because each program is written for a particular student, it is important to have appropriate assessment data available that indicate the student's present level(s) of performance. Areas of assessment would include intellectual and social development, and physical capabilities such as the use of legs, arms, eyes, ears, and speech. The student's age, grade, and degree of learning to date must be considered when setting goals. Equally important are the student's strengths and weaknesses. These would include such things as general health factors, special talents, best mode of learning, and sensory and perceptual functioning.

Information can come from tests given by psychologists, educational diagnosticians, teachers, or others who have worked with the student, or it can come from teacher or parent observations.

Setting Priorities

By looking at the student's present level of functioning, parents and teachers can begin to see critical areas that need attention. These areas can be be pinpointed by having parents, teachers, and the student, if possible, state what they think is most important. These become the high priority learning items. Other areas where weaknesses exist can then be identified.

As the IEP is developed, placement needs become apparent. There must be some correspondence between the number and level of the annual goals set and the amount of time available for instruction. Planners need to consider whether goals can be met within the regular program with consultation for the teacher, with a few hours a week of supplementary instruction, or with more hours of direct instruction by a specialist.

The IEP should be a flexible program (a working document) and not an unchangeable contract. It is a working agreement between professionals and parents working together to help the student progress and gain skills. It is open to review at any time upon request by any of the individuals participating in its design, but it must be reviewed by law at least annually.

Setting Goals

Annual goals can only be the group's best estimate of what the student will be able to do within one year. If goals are accomplished sooner than anticipated, additional goals will be set. There must be a relationship between the annual goals set and the student's present level of performance. The support needed to achieve the annual goals must be documented and the person(s) responsible for such support should be listed.

While the annual goals for each student are established by the planners themselves, the short-term objectives can be obtained from a variety of published sources. A curriculum guide can often be the best tool to use when pinpointing behaviors and sequencing short-term objectives.

Developmental Model

The developmental model is based on the belief that all individuals are capable of continuous growth throughout their lives. This growth includes mastering increasingly complex behavior, attaining greater control over the environment, and enhancing personal strengths and assets. The framework of the developmental model is through effective IEP program planning for special needs students so that the individual's strengths and needs are matched appropriately. A handicapped learner will have developmental delays related to his or her prior degree of isolation or lack of experience. The learner will only bridge the gaps and delays through a focused, structured course with established goals and the opportunity for active participation.

Growth must take place in an atmosphere of human dignity and respect. It involves an understanding of the uniqueness of the student consumer and confidence in his or her ability to take the risks necessary for growth. The right to fail and the support necessary to turn an unsuccessful experience into a means for learning are important aspects of this process.

VOCATIONAL EVALUATION

In developing the IEP with technology components, several questions are bound to arise:

"What interest does the student have in this curriculum?"
"Which classes best match the student's aptitude?"
"Where will the student need assistance or preparation?"

A vocational evaluation keyed to the existing technology courses and placement options within the school and community can provide answers.

Traditionally, vocational evaluations were a diagnostic service associated with vocational rehabilitation agencies. These services were designed to assess a client's specific vocational aptitude and psychological needs, and to recommend a training program. Later this narrow occupational definition was broadened to serve citizens with a wider range of disabilities. Assessment instruments were broadened to test persons without reading or communication skills and content was broadened to include community and daily living skills (mobility, money uses, social maturity, etc.). The emphasis also began to shift toward situational or on-the-job evaluation components. As this service moved into the school systems, the scope of assessment was further extended. In addition to information useful in vocational training and placement, information relevant to placements in school curriculum has become essential. This includes related curriculum offerings that may involve leisure and recreational skills, as well as specific academic needs.

A traditional vocational evaluation process and its product will be described with emphasis on technology education testing. It is up to the technology and special education staffs to determine how to best bring the assessment process into congruence with the specific needs of their school. In its simplest form, the evaluation process should include an assessment of the student's interests, specific and general vocational aptitudes, and work habits and interpersonal skills. The evaluation product is the report and it should include recommendations for specific class enrollment, personal adjustment needs (when warranted), and support for related learning needs within the larger school system.

Rationale for Vocational Evaluation in Public Schools

Currently, learning disabled and educable mentally retarded students who are mildly handicapped make up more than 65 percent of the students classified as handicapped in a typical secondary school. Of the total students enrolled in secondary vocational education programs for the fiscal year 1977 only 2.1 percent were identified as handicapped, according to the Bureau of Education for the Handicapped.

This lack of enrollment is reflected in another set of disappointing statistics. Of the 30 million Americans who are disabled, of whom 11 million are considered employable, only 4.1 million are employed according to the Bureau of Education for the Handicapped. Of those employed, 85 percent earn less than $7,000 annually, while 52 percent earn less than $2,000 annually. These figures stand against the backdrop of numerous studies and projects that have demonstrated that the handicapped are employable and have good work records when compared with their nonhandicapped counterparts.

There are two possible explanations for the statistics of ability and distribution versus the statistics of isolation and deprivation. First, most students are classified or receive services on the basis of testing that emphasizes academic and intellectual achievement, and social and psychological development. Thus, a student is typically placed in special education without any relevant data concerning vocational/technical interests or aptitudes. Instead, referral data often emphasize the need for academic instruction and behavioral programming. Secondly, teacher preparation programs for special educators tend to stress remediation instructional techniques and behavioral methods with very little practical guidance in career or vocational instruction. It is, therefore, not surprising to find that the concept of special education's view of least restrictive environments emphasizes traditional academic and required classes. The same situation confronts the student. How can a student articulate his or her desire to participate in a technology class when he or she knows little or nothing about it? A vocational evaluation service can attempt to correct these difficulties.

The vocational evaluation typically offers the learner a chance to catch up and sample a variety of curricular materials in order to find out what he or she is good at and likes. This information is then systematically collected and summarized so it can be utilized in the development of the IEP. The evaluation not only serves to fill this gap in student programming, but it fulfills one of the legal mandates of the Education for all Handicapped Children Act (Public Law 94-142) and the related Vocational Education legislation 94-482. This legislation mandates a vocational component in the student's IEP as well as the involvement of regular teachers in its implementation.

Vocational Evaluation Process

Most schools that provide a vocational evaluation service base it on a test center that a student attends from ten to thirty hours. This type of center constitutes a *formal* or *clinical* approach to vocational evaluation. A clinical/formal test center has many advantages. It is efficient, easily organized, and provides for routine observations and standard evaluation procedures. its major limitation is the lack of realism. Thus, the most thorough evaluation process also incorporates an *informal* or *situational* assessment. In this instance, data is collected from teachers or employers about the student's performance in an actual community or class setting. The advantage of this type of evaluation is the realism that can lead to successful predictions of future success. How the observations can be standardized and how they can be systematically collected for purposes of summarization and dissemination becomes a concern. A comprehensive evaluation system can emerge from the two components: formal and informal. One setting emphasizes global testing and the other specific training and each can complement the other.

Formal/Clinical Evaluation Systems

Testing centers normally employ three types of test instruments: paper and pencil tests; standardized vocational aptitude and perceptual tests; and work/job/class samples. There are some drawbacks to this type of vocational evaluation system that all professionals should be aware of. The most serious one to consider is the fact that:

> [The student] comes to the evaluation as a fellow human being—a human being who is often fragmented and incongruent. . . . Most of the current vocational evaluation programs are designed to further fragment its clients so that the test scores can be (easily) obtained, objective observations can be recorded, and performance records can be maintained. The client often becomes lost in the maze of unrelated, multidirectional techniques and his performance on these techniques lends support to the fact that he is an incongruent, fragmented individual who is in need of adjustment services (Baker and Sawyer 1971, p. 59).

Another weakness should be considered. Because these tests are highly specialized, results often relate to occupational cluster traits rather than to educational programs. This makes interpretation and application of results sometimes difficult. This situation can be improved by adhering to the following principles in the selection, design, or adaptation of test instruments:

> No one test, especially standardized paper and pencil or vocational aptitude tests, should become the sole basis of a vocational evaluation recommendation.
>
> Tests must be adapted to accommodate learners with communication limitations so that tests are not inadvertently testing communication abilities (e.g., adapt tests and administration of tests for nonreaders, hearing or seeing impaired, persons with limited written skills).
>
> The tests selected must have relevancy for placement in school programs. This implies that much of the testing should have content taken from or based on the curricular offerings as well as local occupational opportunities.

Paper and pencil tests

Both information-gathering and standard written tests can be classified as paper and pencil tests. Areas often assessed with this method include:

- *Intake (Initial) Interviews:* Data is gathered on personal history, outlook, and interests, and the student is oriented to the evaluation process.
- *Intellectual/academic skills.* Data is gathered on intellectual achievement and ability in basic academics like reading and math.

- *Independent living skills.* Data is gathered on ability to use money, maps, and schedules, though this is best done in a situational test.
- *Occupational and career interests.* Data is obtained from student's indirect responses regarding values and lifestyle goals or more directly from student's identifying pictures of people engaged in different career-related activities.

It is helpful to begin with interest inventories because they assist the evaluator in selecting those test instruments that will be used later. Examples of standardized tests are:

AAMD-Becker Reading Free Vocational Interest Inventory. American Association of Mental Deficiency, 5201 Connecticut Avenue NW., Washington, DC 20015.

Geist Picture Inventory. Western Psychological Services, 12032 Wilshire Blvd., Los Angeles, CA 90025.

Wide Range Interest and Opinion Test. Jastak Associates, Inc., 1526 Gilpin Avenue, Wilmington, DE 19806.

Social and Prevocational Information Battery. CTB/McGraw Hill, Del Monte Research Park, Monterey, CA 93940.

Kuder Inventories. Science Research Associates, 155 N. Wackee Drive, Chicago, IL 60606.

The technology instructor can add to this informal interest test. Such a test might help pinpoint areas of interest *within* the technology curriculum. The test could contain a series of pictures showing different processes and activities representative of the various technology classes available. Additionally some "distractor" photos from related hands-on activities (such as art and science) might also be included. With an equal number of pictures randomly dispersed from each area, the student will most often indicate areas of high strength by choosing one or more areas over the other. This is evaluated simply by counting the area which was most often chosen.

Paper and pencil tests lack realism. They are often the least interesting or motivating for the student. These tests normally comprise a small part of the beginning of the evaluation process.

Standardized Vocational Aptitude and Perceptual Tests

When a test is standardized, norms have usually been established that allow the evaluator to compare the student's performance with so-called averages to determine the student's strength or weakness in the tested area. Standardized tests have come under attack recently, especially when applied to disadvantaged and handicapped students. Used as indicators, and not determinants of evaluation recommendations, they can play a useful role. Many tests provide tables of norms for specific populations (e.g., vocational students, college students, handicapped sutdents, age/grade, etc.).

Tests of perceptual skills are often tests of isolated perceptual aptitudes. Examples of these types of tests are:

Perceptual-Visual Figure Ground. Western Psychological Services, 12031 Wilshire Blvd., Los Angeles 90029.

Kinesthesis. Western Psychological Services.

Auditory Sequential Memory. ITPA, University of Illinois Press, Urbana, IL 61801.

Spatial Concepts (Directionality). Hope for Retarded Children and Adults, San Jose, CA.

Form Perception. Address Unknown. Bender Visual Motor Gestalt. Address Unknown.

Because of their unusual nature, a demonstration of these tests in necessary for complete understanding. The aptitudes tested include the sense of touch (kinesthesia); the ability to perceive forms or different shapes (form perception); the ability to learn or memorize information in visual or hearing modes; visual acuity and abilities; the ability to plan work and anticipate the moves (motor planning).

Standardized vocational tests check motor coordination and dexterity. Some test the ability to use tools and test motor coordination from fine through gross (whole body). Many of these tests are used by industry as part of the initial screening process for applicants. Most of these tests are timed and the errors are counted. Typical tests include:

Purdue Pegboard. J. A. Preseton Corporation, 71 Fifth Avenue, New York, NY 10013.

Bennet Hand Tool Dexterity Tests. Psychological Corporation, 757 3rd Avenue, New York, NY 10017.

Crawford Small Parts Dexterity Tests. Psychological Corporation.

Stromberg Dexterity Test. Psychological Corporation.

Pennsylvania BiManual Worksample. American Guidance Service, Publishers Bldg., Circle Pines, MN 55014.

Valpar Eye-Hand-Foot Coordination. Valpar Corporation, 3801 E. 34th Street, Tucson, AZ 85713.

Information from these tests rarely serves as an indicator for predicting success in a specific technology class. The tests can be used to determine the instructional modes best suited for the student, or to pinpoint causes for difficulty on work/job/class samples. The technology teacher can add to the repertoire of test materials used to assess student aptitudes related to the teacher's area. Examples of these additions might include:

Tests of the ability to measure or read numbers on measuring instruments.

Tests related to the ability to use blueprints or match them to objects.

Form perception (e.g. the ability to identify different materials, finishes, welds).

Fine motor coordination using tools and materials from technology labs.

Work/Job/Class Sample

Work/job/class samples should be used for gathering 90 percent of the evaluation data with the other two components supplementing sample findings or used to help plan the evaluation of the individual student. Samples are hands-on test instruments for the most part. Generally, these instruments use realistic materials, tools, and equipment. As simulations of actual class or job conditions, they contain what is termed "face validity." Samples are time-consuming to create as well as to deliver. However, they tend to yield the most useful types of information when compared with paper and pencil, and standardized tests. Teacher-made samples based on existing local job or class options can become the basis of an evaluation system. A curriculum analysis should become the basis for constructing a sample related to a technology class. The sample should be self-contained and allow as much independent student administration as possible to make them efficient test instruments.

There are two basic approaches to the design of a work/job/class sample. One design is based on an analysis of a specific job or class (e.g., auto parts counter sales or welding class). The other is based on clustering several skills from several different jobs or classes to form an occupational cluster or work sample. In this example, several skills from the mechanical cluster may be grouped to reflect the total curriculum (e.g., body work, engine, electrical, suspension). Or they can be grouped by the type of skills necessary to succeed in the technology classes (e.g., fine and gross motor coordination, measuring skills, reading technical literature, use of worksheets). A curriculum analysis leads to a listing of possible prerequisite skills related to the class(es). Examples of curriculum analysis and prerequisite lists are found in chapter 5. The final result should incorporate actual materials, processes, and tools to assess these skills. For example, charts related to the use of welding rods, identification of the parts of a gas welding torch, and safety information, and consequent tests can all be used along with sample welds, which the student can match to descriptions of acceptable welding beads. The sample can suggest that the student have a daylong welding experience in the welding lab to fully assess the student's interests.

There are several different ways to write a work sample. Located at the University of Wisconsin-Stout is the national center for collection and dissemination of work samples. One can obtain a list of work samples already developed and available for duplication by contacting

Information Services
Materials Development Center
University of Wisconsin-Stout
Menomonie, WI 54751.

Additionally, it might be prudent to adopt the style recognized by Stout for the design of teacher-made work samples. The Materials Development Center (MDC) spells out this style in their *Work Sample Manual Format*.

Commercial products exist that are based on a work sample approach. However, one of their drawbacks is the lack of correspondence to local community work and school placement options. Likewise, the tests are more generalized in design and in the consequent data provided. These products can augment and round out a fledgling teacher-made vocational evaluation system, however. Available products include:

Broadhead-Garret Vocational Skills Assessment and Development Program. Broadhead-Garret, 4560 E. 71 st Street, Cleveland, OH 44105

Comprehensive Occupational Assessment and Training System (COATS). PREP, Inc., 1575 Parkway Avenue, Trenton, NJ 08628

Hester Evaluation System, 120 S. Ashland Blvd., Chicago IL 60607

Micro-TOWER. ICD Rehabilitation and Research Center, 340 E. 24th Street, New York, NY 10010

The Singer Vocational Evaluation System. Singer Education Division, Education Systems, 3750 Monroe Avenue, Rochester, NY 14603

Talent Assessment Program, TAP, 7015 Colby Avenue, Des Moines, IA 50311

The Valpar Component Work Sample Series. Valper Corporation, 3801 E. 4th St., Tucson, AZ 85713

Vocational Information and Evaluation Work Sample (VIEWS), Vocational Research Institute, 1913 Walnut Street, Philadelphia, PA 19103.

Wide Range Employment Sample Test (WREST)

Tool Tech Today. MIND, 181 Main Street, Norwalk, CT 06851.

Work Habits/Personal Adjustment/Training

Additionally, the way in which the student learns and comports himself or herself can also be observed. Work habits to look for include:

Neatness
Endurance
Clean-up
Acceptance of criticism
Harmonious peer/instructor interaction
Quality (types of errors and corrections)
Safety
Independence
Reaction to stress

By recording the kind of instruction used and the possible time associated with it during the administration of a work sample, as profile on student learning style can be obtained. Sometimes this means purposefully varying instructional techniques for purposes of observation. The events to be observed include:

The type of instruction—verbal, demonstration, written, modeling, exploration, etc.

The time taken to instruct or the number of trials

The ability to recall instruction

Finally, the evaluation should provide the opportunity for the student to bring up personal concerns. Similarly, some students must be counseled during the evaluation if the evaluator feels the student is not performing to full ability due to personal adjustment problems. These, too, become part of the evaluation process.

Informal Evaluation

An informal evaluation can emerge from an existing work experience, a cooperative work program, or a planned student placement. In this last instance an entry-level class would be the logical place to obtain information. The only necessary ingredient is the systematic observation of the student and the collection of this data. This type of program obviously emphasizes actual training over evaluation procedures. It requires additional input from the teacher or employer. Examples of instruments used to collect work-behavior data are found on pages 33 and 34.

To collect information on specific class vocational skills, observation sheets similar to figures 5-1 and 5-2 can be used. Another approach involves use of a pretest from a specific class. Pretest content or instruments can simply be adapted from existing curricular materials. Likewise, the test procedure can be based on individualized demonstrations that are standardized in format, followed by timings of the student's performance, counting of errors, or both. An excellent example of this approach is the Vocational Education Readiness Test (VERT) developed at Mississippi State University. VERT is used to assist local school districts in developing and implementing an inexpensive vocational assessment program ranging in cost from $500 to $1,000 for materials. It is based on directed instruction followed by observation. See figures 2-1 through 2-5 on pages 35, 36 and 37.

The advantage of such a program is that it is indexed to existing secondary vocational education trade programs instead of general occupational (Dictionary of Occupational Titles referenced) areas as found in commerical products. Copies of the VERT manual can be obtained from:

Research and Curriculum Unit for Vocational-Technical Education
P.O. Drawer DX
Mississippi State, MS 39762

COOPERATIVE WORK EXPERIENCE PROGRESS REPORT

Business Name ______________

Business Phone ______________

Employer's Name ______________

On the Job Evaluator ______________

Dates: from ______ to ______

Student's Name ______________

Student's Home/Work Phone ______________

School Personnel ______________

School Phone ______________

Daily Work Schedule Time: ______________

From: ______________

To: ______________

Transportation: ______________

SUPERVISOR: Please rate your trainee by checking the appropriate circles below:

APPEARANCE (General impression given to public)	***Excellent*** O Always neat and attractive	***Good*** O Clean and presentable	***Poor*** O Barely presentable	***Unacceptable*** O Unpresentable to public	Unable to rate O
ATTENDANCE AND PUNCTUALITY	***Excellent*** O Always on time; prearranged absences	***Good*** O Usually dependable	***Poor*** O Produces a low volume of work with many mistakes	***Unacceptable*** O Habitual absences and latenesses	Unable to rate O
QUANTITY OF WORK	***Excellent*** O Turns out large volume work	***Good*** O Turns out adequate volume work	***Poor*** O Produces a low volume of work	***Unacceptable*** O Below average volume	Unable to rate
QUALITY OF WORK	***Excellent*** O High degree of accuracy	***Good*** O Rarely makes mistakes	***Poor*** O Many mistakes	***Unacceptable*** O Frequent mistakes	Unable to rate
ATTITUDE (General attitude; willingness to cooperate)	***Excellent*** O Always cooperative and eager to do a good job	***Good*** O Enthusiastic; cooperative and willing to learn	***Poor*** O Reluctant to do more than has to	***Unacceptable*** O Uncooperative; puts out little effort	Unable to rate O
INITIATIVE (Amount of supervision and instruction required)	***Excellent*** O Eagerly seeks work; always carries out assignments	***Good*** O Works well with a minimum of supervision	***Poor*** O Must be told what to do; heavy supervision	***Unacceptable*** O Waits to be told what to do; always needs supervision	Unable to rate O
HUMAN RELATIONS (How employee gets along with coworkers, supervisors and public)	***Excellent*** O Gets along very well; adept at handling delicate situations	***Good*** O Gets along well; rarely has personality problems	***Poor*** O Occasionally causes trouble or friction	***Unacceptable*** O Consistently causes trouble; endangers customer good will	Unable to rate O
SAFETY/ INDEPENDENCE (How employee works, how safe is employee)	***Excellent*** O a. *Reports* injuries accidents/damage - *always* b. *Sets up* materials/equipment c. Follows *procedures 100% time* d. *Always asks* when unsure c. *Returns* supplies and cleans up exactly as instructed	***Good*** O a. Reports - *always* b. Sets up but *must be checked* c. Follows procedures 80% time d. *Sometimes* asks when unsure e. *Returns* supplies/ cleans - some errors	***Poor*** O a. Reports - *Never* b. Sets up - *supervision required* c. Follows procedures *only with supervision* d. *Does not ask* when unsure e. Supplies returned and cleans sloppily	***Unacceptable*** O a. Reports - *Never* b. Sets up - *Never* c. Horseplay, unsafe: accident/injury/ damage occurs d. *Does not ask* when unsure e. *Needs continual* supervision	Unable to rate O

Job skills student performs best:	Examples where student needs improvement:

Progress Report adapted from Portland Public Schools, Work Experience Report Form

SAMPLE EVALUATION SHEET

Student ______________________ Job ______________ Supervisor ______________ Date ________

Notes/Job Description: Weekly Percent ________

Work Habits / Work Skills	Criteria		Data							Comments and Observations
	X - Not Observed (See Comments)	? - Unsure (See Comments)	Daily Tally					Weekly		
	I - Acceptable/Competitive	0 - Change Needed	M	T	W	T	F	%	Avg.	
I. ATTENDANCE 1. Daily										
2. Notifies Supervisor										
3. Punctuality	Within ______ minutes follows schedules	___, ___, ___, ___, ___ (daily nonpunctual)								
II. PERSONAL APPEARANCE 4. Hygiene	Acceptable	Offensive								
5. Grooming	Acceptable-hair, skin	Offensive/Noticeable								
6. Dress	Appropriate, clean, neat	Inappropriate, dirty, unkempt								
7. Posture	Erect, alert, good	Slovenly, tense, poor								
8. Other										
III. PERSONAL ATTITUDES 9. Self-Control	Even tempered, controls anger, frustration, accepts changes	Moody, sullen, displays emotions, easily upset, resists change								
10. Self-Image	Confident, cheerful, positive, enthusiastic, praises others	Puts self in negative terms, mal-content (job dissatisfaction)								
11. Interpersonal Skills (Peers/Co-Workers)	Tactful, diplomatic, sensitive, at ease in groups, participates in group or conversation, is sought out by others, sense of humor, handles teasing	Blunt, arrogant, selfish, domineering, shy or anxious in a group, stays away on breaks, brittle, easily set up, irritates others, teases others, gossips								
12. Cooperation	Helps out, extra effort, pride in group effort, compromise	Must be asked, does only what is expected, critical of others, no compromise								
13. Other										
IV. SUPERVISION 14. Mature Work Relations	Accepts supervisor; at ease, asks for help, greets them, appropriate conversation, pro-fessional image (distant)	Critical of authority, tense, confused, slow or lack of eye contact in their presence, too personal, gossip, seeks attention/contact (too close), fawning								
15. Initiative/Independent	Seeks out new work, uses time wisely, rarely idle, tries to solve problems, works on own without supervision	Wastes time, often idle, needs to be told too often how/what to do (prompts)								
16. Follows Rules	Listens and accepts rules. follows procedures and carries out instructions to best of ability	Challenges or tries to change rules and procedures (bargains, trades, exceptions), disregards rules or procedures or instructions								

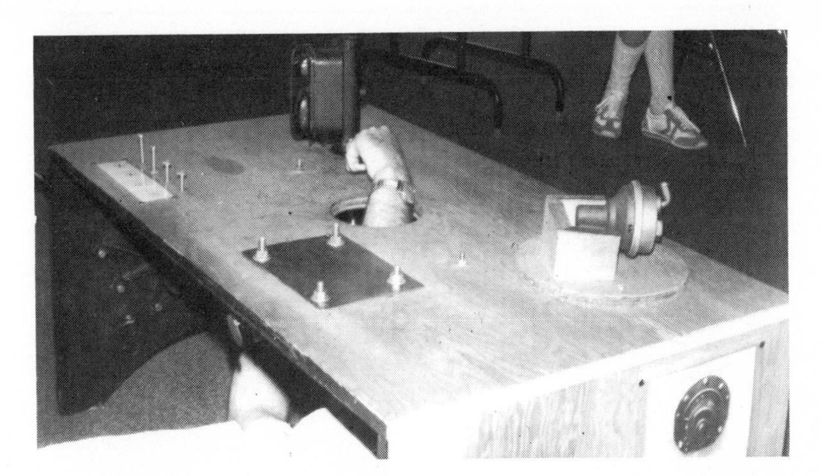

Figure 2–1. *Preliminary exercise: The evaluator will identify the 0/16" open-end wrench and the 9/16" socket and ratchet to be used. The evaluator will then demonstrate how to remove and replace the brake master cylinder.*

Given the auto mechanics mock-up, the student will remove and replace the brake master cylinder using the 9/16" socket with ratchet on the bottom while in the supine position. Note time. (Courtesy of Vocational Education Readiness Tests (VERT), Research and Curriculum Unit for Vocational-Technical Education, Mississippi State University, MS 39762)

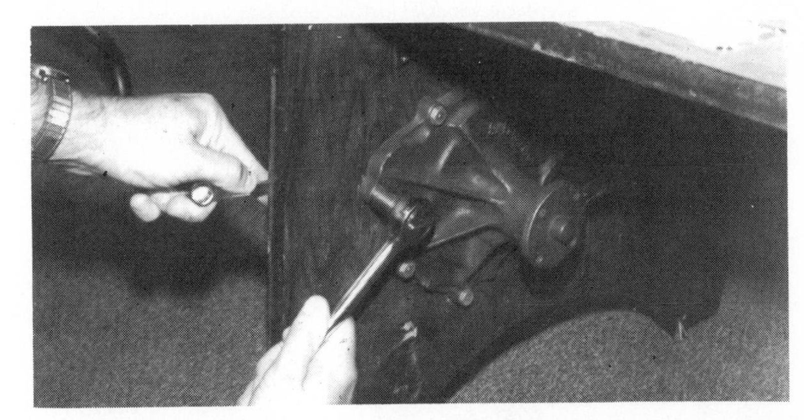

Figure 2–2. *Preliminary exercise: The evaluator will identify the 1/2" open-wrench, and the 1/2" socket and ratchet to be used. The evaluator will then demonstrate how to remove and replace the water pump. (Special instructions: The evaluator will demonstrate how to reverse the ratchet but leave it set for student to begin the task. Instruct student to tighten the bolts only to the degree that they cannot be loosened with the fingers. Instruct student to completely remove the water pump from the mock-up, place it on the floor with all bolts removed, and then replace it. Have student attach socket and extension as part of this task. In addition, the evaluator should demonstrate the correct body position to be used for removal of the water pump.)*

Given the auto mechanics mock-up, the student will remove and replace the water pump using the 1/2" open-end wrench in the left hand and the 1/2" socket with ratchet with extension in the right hand while lying on the right shoulder. Note time. (Courtesy of VFRT.)

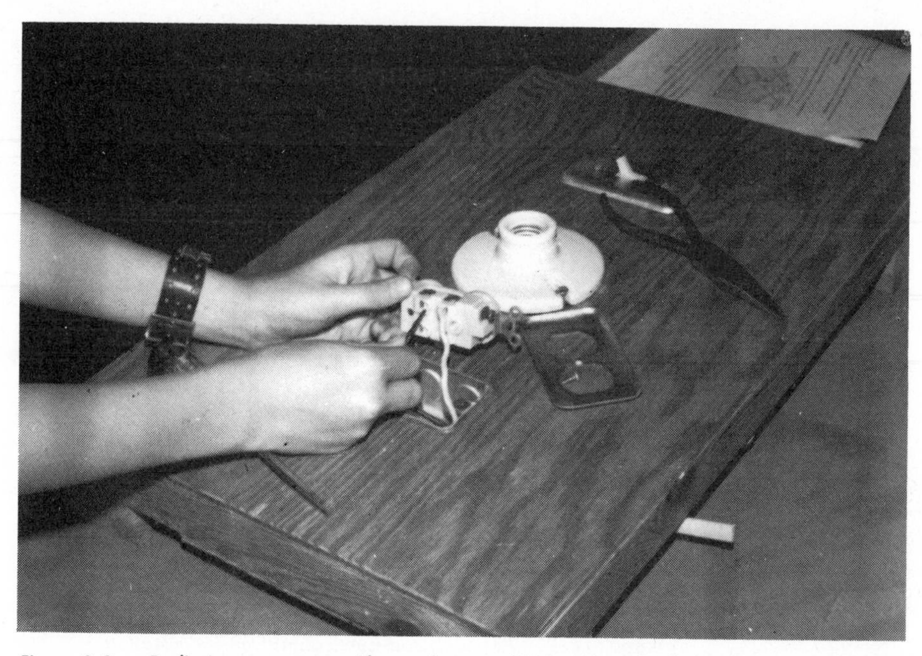

Figure 2–3. *Preliminary exercise: The evaluator will identify a duplex wall receptacle cover and outlet box, and will demonstrate how to attach the wires to the quick-wire receptacle. (Special instructions: Use a standard 3/16" screwdriver for cover and a standard 1/4" screwdriver for attaching receptable to outlet box.)*

Given a duplex wall receptacle, cover outlet box, #12/2 cable, and a 3/16" and a 1/4" screwdriver, the student will insert the white wire to the white wire opening on the quick-wire wall receptacle, the black wire to the other quick-wire opening, attach the receptacle to the outlet box, and attach cover. Note time. (Courtesy of VERT.)

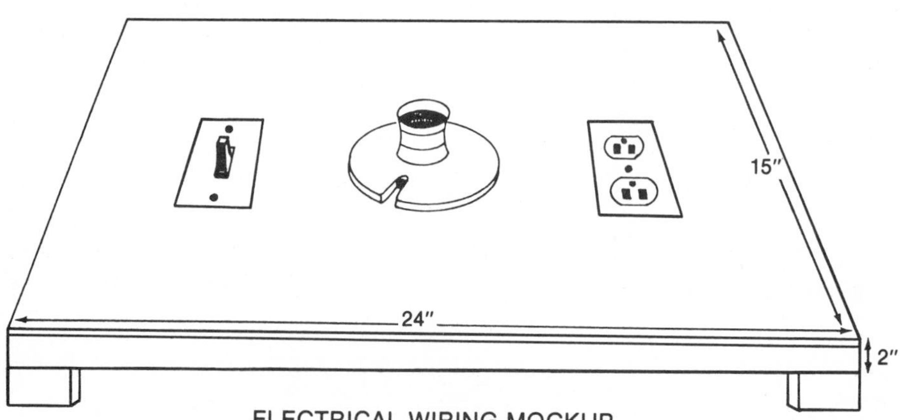

Figure 2–4. *Electrical wiring mock-up. (Courtesy of VERT.)*

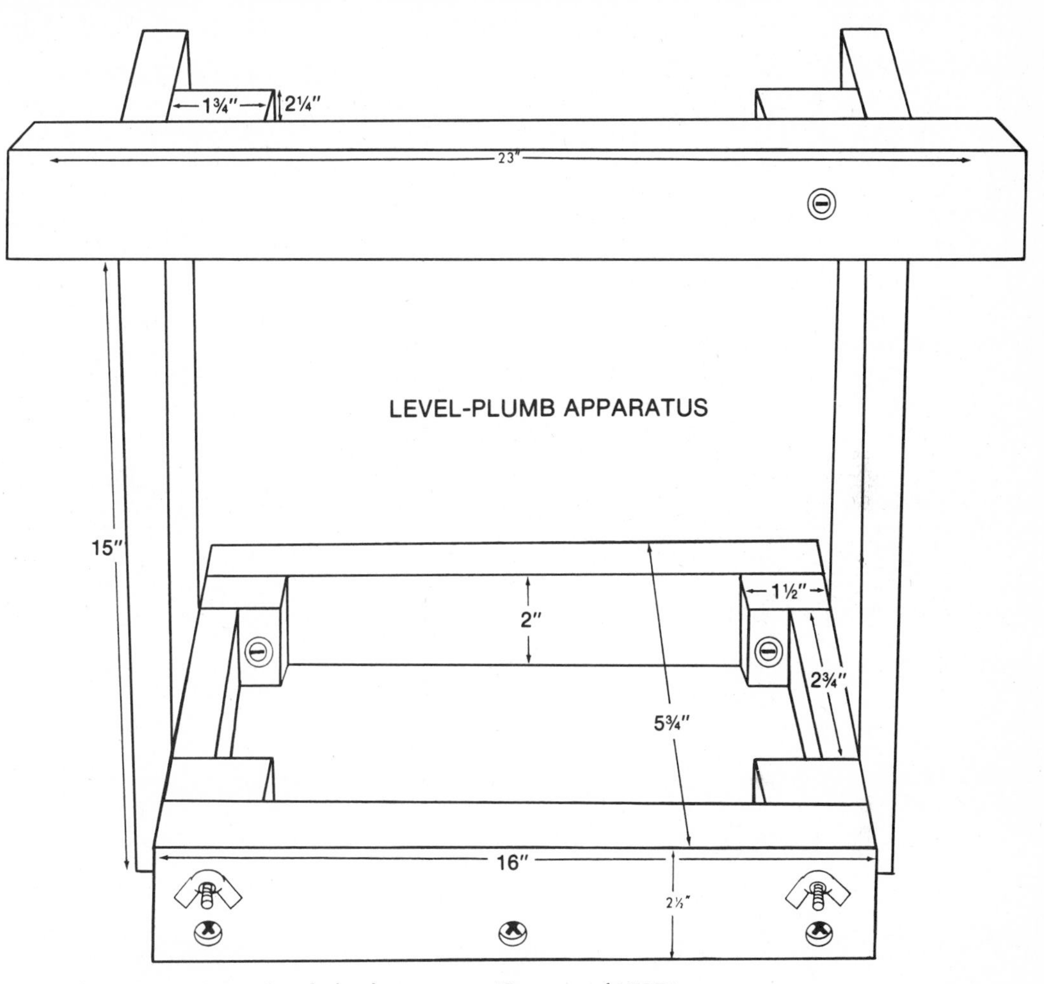

Figure 2-5. Level-plumb apparatus. (Courtesy of VERT.)

Evaluation Product

In conclusion, four qualities should summarize the evaluation process and product:

1. The process should systematically gather information on a student's performance.
2. The evaluation must be comprehensive.
3. The result should comprise formal and informal elements and not rely soley on a clinical two-week evaluation.

VOCATIONAL EVALUATION REPORT

I. Demographic information
   A. Student name
   B. Date of birth
   C. Referral source
   D. Period of evaluation
   E. Report written by
   F. Evaluators

II. Background information
   A. Reason for referral
   B. Student's interests
   C. Present academic performance or programming

III. Vocational aptitudes summary
   A. Dexterity
   B. Gross motor and physical capacities
   C. Perceptual skills
   D. Recommended training style
   E. Work habits and personal behaviors
   F. Community/independent living skills

IV. Recommendations
   A. Summary of vocational training
   B. Recommendation for vocational training
      1. List classes or programs
      2. List type of placement suggested
      3. List related support needs
   C. Recommendation for related academics
      1. List remediation or compensation suggestions
      2. List suggestions for extending present skills
      3. Describe how to adapt current academic programs to life-centered program as warranted
   D. Recommendation for specific personal adjustment needs
   E. Recommendation for community living needs
      1. Leisure/recreation
      2. Mobility/travel
      3. Home/residential living
      4. Consumer/citizen/career awareness
      5. Special programs

4. The report should provide specific recommendations for placement as well as summarize test results in language that can be understood by teachers and parents.

Information can be gathered from a variety of sources at any time, including the school nurse, counselors, special and regular teachers, employers, parents, vocational evaluators, and the student. Summarizing the information from these sources is critical if it is to guide the development of the student's IEP. Summarization should be an on-going process similar to IEP reviews. After a formal report is written, following a clinical evaluation, additional experiences can be logged on forms for summary purposes. An example of this might relate to the cooperative instructional agreement found in chapter 5. An outline for a formal vocational evaluation report is shown on page 38. (See Valpar-Spective n.d.; Mason and Bentley 1973 for other examples.)

STUDENT PLACEMENT OPTIONS

The Life-Centered Technology Education Model discussed in chapter 1 is based on a continuum of educational services or class offerings that leads toward the least restrictive class setting. This is embodied in the movement from an entry-level/special education class to an exploratory and cluster level/regular class placement. Following a vocational evaluation, or preceding a cooperative instructional agreement (see chapter 5) that leads to a regular class placement, the major focus evolves from curriculum content toward individual student needs. This is reflected in the development of the IEP as the special education teacher considers the specific needs of the learner when placed within the existing curriculum options. The existing curriculum options represented in the Life-Centered Technology Model can be expanded when considering the placement options for the individual student. These placement options are listed on page 40.

Starting with the most restrictive and most specialized entry-level program, the setting becomes self-contained and highly supervised. This pre-entry level class would be for students who may have work habits or personal behaviors that would not be acceptable in a regular class. This instructional setting would constitute a work adjustment program with individual behavior management programs written for each student. Students might work individually or in crews or groups. In either event, the technology serves as the basis of the activity around which instructors provide the student with feedback and guidance concerning behaviors. For example, a student might be placed in this class to learn how to control his or her temper, how to cooperate with others, to improve his or her attendance. A student might also be placed here for instruction in basic skills. A student might be learning how to use a hand tool for the very first time, how to write his or her name, or how to tell time, which could relate to some

STUDENT PLACEMENT OPTIONS

Most Restrictive ——————————————→ **Least Restrictive**

Least Specialized ——————→ **Most Restrictive**

Entry-Level Programs

Pre-entry for individual Remediation

Work and personal adjustment
- Individual behavior management
- Counseling
- Group Projects

Specific Skill Training
- Motor/coordination training
- Perceptual training
- Basic academics/communications

Entry training for regular class placements

Entry-level program

Adapted exploratory program
- Team teaching
- Individualized methods

Regular Exploratory/Cluster Level Programs

Modification of regular placements

Placements with work or personal adjustment goals
- Goals regarding work habits
- Goals regarding personal behaviors
- Goals regarding socialization/motivation

Placement with specific skill training goals
- Partial class placement
  - regular course but focus is on mastery of part of curriculum over extended time

Specialized placement
- regular course setting but content is geared to specific individual needs

Regular class placements with support

Supplemental program offered in resource room
- Academic, tutorials, alternative assignments

Supplemental program in regular class
- Tutorial assistance
- Team Teaching or Planning
- Modification of instructional setting, instruction, requirements, or grading

technological process. Another student might be a nonreader or nonverbal. In this instance the class would be used to establish a basis for communicating (e.g., respond to directions like "on/off," "apply more/less pressure," "clean up"), which can be used in the technology lab. Another student might need to develop an alternative communication system that might involve sign language, language cards, and other expressive modes built around a technical vocabulary. A nonreader might begin by building a sight vocabulary related to safety and tool identification.

The next movement might be within a restrictive setting but toward a less specialized class, such as entry training. In this instance the instructor wants to test out the growth and development of the student in a more controlled setting. For example, when unsure of the student's ability to maintain control over temper, or the practicality of a specific language card system, the entry training class serves as a proving ground. Some of these classes may be integrated with nonhandicapped students so that an aspect of normalization takes place.

The other option to consider is a specialized placement within a regular class; this would constitute a modification of regular placement. In one instance a student could be placed in a regular class with a simple modification that sustains his or her growth in personal adjustment. For example, the student might be placed in the auto class and receive half his or her grade based on the work performed and the other half on specific work habits or personal behavior improvement or maintenance (e.g., attendance, contributing to personnel system, etc.). In another case, a student might be placed within the regular class for specific skill acquisition. In one instance it might mean learning just one unit within the course, such as lubrication, but it may take place over the entire semester. Another student might be placed in the regular class but with specialized goals that are not immediately related to the course content. For example, placement in wood clusters for the purpose of building strength and coordination with hand tools, or placement in electronics clusters for practice with small tools and parts to build fine motor coordination. In these last examples, the principle of normalization is being implemented. That is, the student might be a senior who will benefit from interaction with his peers and likewise the stimulation of a realistic environment. The technology lab then becomes a resource lab serving the needs of all students who can benefit from some type of interaction with technology and people. A student having difficulty with math may completely change his attitude and consequent learning upon finding a useful application of math in the technology classes.

The least restrictive and least specialized class is the regular class with supplemental components. Some students will want or need supplemental assistance back in the resource room. This may be tutorial help in reading the texts, reading and answering tests, filling out worksheets, or studying and practicing specific skills. Other students might desire and benefit from assistance directly in the lab setting itself. In this case the education specialist can assist directly, help set up a peer tutoring system, or help the regular instructor to adapt the learning environment.

RELATING THE IEP TO TECHNOLOGY EDUCATION

How can the technology instructor insure that the IEP process and related conference will reflect the needs of the student in the lab setting? The IEP conference can set in motion the coordination needed between parents, students, and instructors. The IEP conference is not an end itself and will not eliminate the need for further planning and cooperation regarding the day-to-day modifications that may be warranted. The overall purpose of the IEP conference is to insure that the learner achieves maximum participation in the least restrictive environment by planning a program that outlines support services related to placement. The IEP conference covers the entire program for the student. It is possible that some students will need no further support in the technology program, in which case the IEP conference becomes merely an introductory meeting. For another student, the entire conference may focus on the concerns of the parents, teachers, and student as regards the technology program because it may be the only new element in the student's program.

What types of information will the teacher learn at the IEP conference? Primarily, the regular teacher will learn a great deal about the prospective student. Topics normally covered include:

1. Previous work in technology or vocational education.
2. Career goals that relate to the classes being considered.
3. Specific impairments of the student that relate to the classes and that might become handicaps.
4. Rationale for placement in a given course. For example, is the student being placed for very specific reasons with individual goals or is it an experiment to assess the student's interests and abilities? Questions like this help the teacher to adjust his or her expectations.
5. Questions about the course that the student may need to have answered before signing the IEP.

At the IEP conference the technology instructor will meet the members of an interdisciplinary team: the director of special education, the vocational evaluator, the counselor, special education teachers, the diagnostician, and the parents. (This list will vary with individual students.) This team can answer questions regarding behavior goals and methods appropriate to the student, types of observations that would be useful to the diagnostician or special education teacher, parental concerns and insights, the types of tangible support special education can furnish, and the types of perceptions others have of the technology curriculum. As long-term goals are hammered out, the entire team can assist in stating the goal in terms that reflect the student's current abilities. At this point the technology instructor can better describe the logistics of the lab setting, additional instructional support that will make the goals realistic, his or her educational background related to special needs learners, and any concerns he or she may have. As a general guideline, the long-term goals should be kept general and simple.

Each school district has its own IEP format and forms related to the process. Shown here is a generalized IEP form that may prove useful if the school's current forms do not adequately represent the placement needs in a technology setting. This form is adapted from one used at the Owen-Sabin Occupational Skills Center in Oregon and is presented for its flexibility and simplicity. Too often, IEP forms using standard task analyzed lists and coded statements look impressive on paper but fail to demonstrate how the placement can actually be carried out. It is unrealistic to state an evaluation technique, specific instructional method, or material regarding each performance objective of the course. However, this form allows the teacher to plan without becoming overly specific—thus allowing for later changes as warranted. Likewise, it is unrealistic to assume that the regular teacher will be able to provide extensive additional data on student performance unless he or she receives support in this area.

GENERAL IEP FOR TECHNOLOGY EDUCATION

Annual Goal

The student will participate in the technology program in the following class:

____

The long-term goal is to receive instruction in the following:

______Career/Occupational Awareness
______Awareness of technology
______A specific technical skill
______Socially appropriate behaviors
______Related academic skills
______Successful work habits: ______________________
______Other: ______________________

Short-term Instructions Objectives

Duration: The student may take longer than normally required but no longer than one year to master the following skills, without reevaluation of the goals and objectives.

Objectives and Evaluation: The student performance objectives are to include but are not limited to the performance objectives listed below. They are to be mastered with a 90 percent proficiency as evaluated by the instructor using regular and/or special techniques. Supporting roles are indicated below.

Accountability
IEP Team

______________________ Date ______________

____

______________________ Annual Review Date ______________

____

Performance Objectives	Teaching Responsibility				Modifications/Comments/Materials		
(Use performance objectives listed for the course as a guide)	Regular Instruction	Tutorial/Team Support	Special Personnel	Other/Comments	Regular Course Objective	Modified Course Objective Describe:	Specialized Materials

Following the IEP conference the next logical move is for the special and technology educators to outline a cooperative instructional agreement that will further specify how the student will be taught. This concept is discussed in chapter 5.

After looking over the IEP form the instructor might ask, "Am I expected to individualize and change everything in my course?" Many technology classes already incorporate some degree of individualized instruction, and may only require minor adaptations to make the course more accessible to special needs learners. Examples of existing individualization common to many technology programs are:

- Competency based evaluation of student skills
- Individualized instructional materials or packages allowing learners to progress at their own rate
- Individual attention from the instructor during the hands-on part of the course

Some of these techniques merely have to be extended in scope through the use of peer tutors, alternative assignments, or performance indicators.

After the student has been placed, the instructor may call for an IEP review in order to ask for greater assistance or changes in the goals and methods. For example, after teaching the student the instructor might notice that the student needs more support. During a review, the role of the special educators in providing support for academic or related needs can be articulated. Also, alternative routes for grading or questions regarding credit may have to be discussed. Finally, the instructor may believe that the student will profit from some different material. Perhaps the student is unable to absorb all the information from lectures regarding welding. The instructor feels that the student can learn welding if an audiovisual self-paced product were purchased to assist in instruction. The IEP review could then attain a consensus for the need of the materials and approach both special education and vocational education personnel with a documented need.

CURRENT PROBLEMS

Mainstreaming has been met by two directly opposing faces. Some professionals describe it as a visionary and progressive concept, while others condemn it as a Pandora's box leading to chaos and confusion. Public Law 94–142, which mandated mainstreaming, actually mandated that a student be educated in the least restrictive educational setting possible. In other words, the goal is to educate handicapped students with nonhandicapped students to the maximum extent possible. It clearly called for an end to segregated curricula, classes, and schools, but it did not rule out specialized programs and support services. It is obvious to most educators, parents, and students that mainstreaming is still in an experimental stage and that the law requires further refinement. Difficulties experienced in mainstreaming have been:

1. Haphazard placement of students in regular curricula
2. Lack of support for teachers in regular curricula
3. Underfunding of the goals mandated by P.L. 94-142
4. Open hostility in the schools

Haphazard placement refers to a student being placed in regular classes without regard to his specific needs, orientation of regular teachers, and the student's past educational experiences. Another difficulty can arise if the students in a regular class have not been fully involved in the integration process. The student with a visible handicap may become the object of ridicule and scorn. Finally, Mary Rita Hanley wrote in *Adrift in the Mainstream* that "special education resource teachers are also often given the responsibility of negotiating with the regular teacher for the placement of special youngsters and their success parallels their negotiating skills." In summarizing the history of special needs learners' involvement in technology, we find a similar haphazard historical precedent. In a survey done of the literature related to industrial education the term "slow learners" is found in the 1920s, in the 1950s one reads about "manual arts therapy," and then the literature proliferates in the 1970s. Regardless of the literature, the author fails to find a comprehensive and generalized model that effectively advocates a blending of two fields in order to serve the life-centered needs of *all* handicapped students. Thus, while experiments flourish, randomness or haphazardness prevails.

The author can cite a personal experience relating to the haphazard placement of students. During roll call on the first day of class, a student did not answer to his name. The author interpreted this as an act of noncompliance. However, upon further investigation he found that the student was deaf and placed in his class without prior orientation.

A second area of concern is the lack of training and support for the regular teacher. Most school's in-service models for regular educators are designed without sufficient teacher input and without regard to the particular needs of individual teachers for instructing special students. Additionally, teachers often have specific questions regarding individual student needs and find it difficult to get meaningful assistance. Another example, from personal experience, occurred in a beginning drafting class. In this instance, the author went to the student's teacher and asked what instructional alternatives the teacher might suggest for a mentally retarded student. The resource teacher responded with, "treat him like all the others." This was insufficient for a beginning drafting class, which is often difficult for all students. In this case, the student's efforts were salvaged by moving him from a regular graded base to a pass-fail one.

Related to these problems are the cutbacks in funding that limits our ability to fulfill the mandates of 94-142. Monies are often not available for the necessary support personnel and programs. In some districts, the "least restrictive" concept has become a way of eliminating special programs and personnel in the name of mainstreaming. For other districts, the decisions regarding future program development has become a bread-and-butter decision between basic academic remediation service verses life-centered curricula.

As a consequence of the previously described obstacles, we find ourselves in the middle of the fourth problem: teacher hostility. The two largest teacher organizations, the National Education Association and the American Federation of Teachers, have viewed the P.L. 94-142 with a cautious acceptance. The skepticism of organizations and the reluctance of regular teachers (and some special education teachers) is understandable. Comments that one may hear include:

"I'll mainstream anyone, if that's all I have to do. I'm already over enrolled."

"What about safety? I can't watch everybody."

"These kids are dumped in my class. I'm not trained as a special ed teacher."

We only have to add these feelings to an already hostile school campus that may be embroiled in conflicts between teachers and administrators over contracts, evaluations, and budgets, where mainstreaming is seen as just fanning the already existing flames. The solution is clear: a program that provides for a "series of planned experiences" and teacher involvement—the blending of two disciplines.

EXPLORATIONS

Activities

Sponsor a debate on mainstreaming with a local special education organization like Council of Exceptional Children (CEC) or Association of Retarded Citizens (ARC).

Design your own IEP form.

Design a work sample.

Design a standardized test that tests student abilities related to a specific technology area.

Obtain samples of vocational evaluations and have the reports explained to you (see Valpar or Stout MDC).

Visit a place that provides vocational evaluations and take standardized tests.

Look at a career interest test or take one yourself. Special interest might be found in a computerized job interest test designed for physically handicapped people. It correlates physical capacities required for a job (as described by a physician—walking, standing, lifting, etc.) with the person's interests and abilities. Refer to the Job-Related Physical Capacities Research Project, Florida International University, Tamiami Campus, Miami, FL 33199.

Readings

Micali, James J., and Scelfo, Joseph L. *Guidelines for Establishing a Vocational Assessment System for the Special Needs Student*. New Brunswick, N.J.: Vocational-Technical Curriculum Laboratory, Rutgers-The State University, April 1978.

Read the magazines cited in the text.

Send for a bibliography from the University of Wisconsin-Stout, Material Development Center. References available include:

> Botterbusch, Karl F. *A Comparison of Commercial Vocational Evaluation Systems*. June 1980.
>
> Botterbusch, Karl F. *Test and Measurements for Vocational Evaluators*. May 1973.
>
> Esser, Thomas J. *Gathering Information for Evaluation Planning*. March 1980.

Read local school district IEPs and policies for referral.

REFERENCES

Baker, Richard J., and Sawyer, Horace W. *Guidelines for the Development of Adjustment Services in Rehabilitation*. Auburn, Ala.: Rehabilitation Services Education, Department of Vocational and Adult Education, Auburn University, Sept. 1971.

Block, Patricia, ed. *Vocational Education Readiness Test Manual*. Mississippi State, Miss.: Research and Curriculum Unit, Vocational Technical Education, Mississippi State University, 39762, 1978.

Brolin, Donn E., and Kokaska, Charles J. *Career Education for Handicapped Children and Youth*. Columbus, Ohio: Charles E. Merrill, 1979.

Brolin, Donn E. *Vocational Preparation of Retarded Citizens*. Columbus, Ohio: Charles E. Merrill, 1976.

Buffer, James T. *A Review and Synthesis of Research in I. A. For Students with Special Needs*. Columbus, Ohio: Ohio State University, Eric No. Ed. 090–394, 1973.

"Career and Vocational Education for the Handicapped." *The Pointer for all educators and parents of exceptional children*. Vol 26, No 4 Summer 1982.

Geiser, Robert L. "The I.E.P. Dilemma. *The Exceptional Parent*. 9:E 14-6, Aug. 1979.

Gickling, Edward E.; Murphy, Lee C.; and Mallory, Douglas W. "Teachers' Preferences for Resource Services." *Exceptional Children*. 45:442-9, March 1979.

Hanley, Mary Rita. "Adrift in the Mainstream?" *The Exceptional Parent*. 19:E3–E6, August, 1979.

Kimeldorf, Martin. "I want to take shop." In *Towards Least Restrictive Environments*, Edited by S. Wapnick and C. Scanlon. Portland, Ore.: Oregon Council for Exceptional Children, Portland State University, 1979.

Kok, Marilyn, and Parrish, L. H. "How the IEP Helps the Shop Teacher." *School Shop 39:19–21, May 1980.*

Lilly, Stephen M. and Smith, Paula. "Special Education As a Social Movement." *Education Unlimited*. 2:7–11, April 1980.

Mason, Kay, and Bentley, James. *Comprehensive Vocational Evaluation, Adjustment and Placement Program*. Columbus, Ohio: Ohio Department of Mental Health and Mental Retardation, 1975.

Nadler, Barbara, and Shore, Ken. "Individualized Education Programs: A Look at Realities." *Education Unlimited*. 2:30–33, April 1980.

Norman, Michael, and Krause, Donna. "Bureau of Education for the Handicapped Reports." *Education Unlimited*. 2:48, March 1980.

Thiel, Suzanne. Course: Vocational Evaluation, Ed. 507. Portland, Ore.: Portland State University, Fall 1978.

Tice, Walter. "Isn't Anyone Out There Really Listening?" *Education Unlimited*. 1:30–31, Nov. 1979.

Torres, S., ed. *A Primer on Individualized Education Programs for Handicapped Children*, Reston, VA: The Foundation for Exceptional Children, 1977.

Valpar Corporation. "The client report—comments and examples." *Valpar-Spective*. 1:4–19, Tucson, Ariz. This is a newsletter published on an informal basis.

"Vocational Evaluation for Special Needs Students." *The Journal For Vocational Special Needs Education*. Vol. 3, No. 3 Spring 1981. NAVESNP Organization.

Thomas, Edward L. Letter to the author dated July 14, 1980.

Chapter Three

IMPAIRMENTS/ DISABILITIES/ HANDICAPS

> FOR USE BY QUADRIPLEGICS, INVALIDS, ELDERLY, WEAK, AND OTHER HANDICAPPED PERSONS

seems an acceptable phrase above a door or ramp, yet

> FOR USE BY LEFT-HANDED HAY-FEVER SUFFERERS

seems ludicrous in comparison. In Elizabeth Pieper's excellent article *Labels, Language, and Self-Image,* she vividly portrays the manner in which words reflect, and often imprison, our thoughts. Any information regarding disabling conditions perhaps should have before it:

> WARNING: THIS INFORMATION MAY BE HAZARDOUS TO ONE'S CONSCIOUSNESS. LABELING CAN LEAD TO STIGMA OR DISTANCING OF ONE'S SELF FROM FELLOW HUMAN BEINGS.

This view can be summed up by paraphrasing Helen Keller who commented that blindness and deafness were not her greatest problem; it was the way that people treated her as a deaf and blind person. The corollary in teaching is that a core of instructional techniques and strategies are needed by teachers of *all* learners. If the instructor emphasizes those practices that are effective for the teaching situation, then one has only to extrapolate from this core of techniques to provide instruction relevant to *all* learners. By focusing on the similarities rather than on the differences we make integration possible. The case for electronics for the deaf, wood shop instruction for the Mentally Retarded, and the like fails to hold up under careful scrutiny or practice. In the last analysis, the Kansas Association for Retarded Citizens declares the right of all citizens to be unique with individual differences with their motto: "The mentally retarded are just like the rest of us . . . they're different."

Much of the material in this chapter was contributed by Sheldon Maron and Peter Wigmore.

Against this general warning or awareness, this chapter will attempt to address the issue of labeling. Different disabling conditions will be described, along with suggested teaching strategies and materials that may be useful in accommodating special needs. The information will be of a general nature. One cannot assume that because information has been read about sensory impairments that school programs can be written without specific input from the student, assessment of the student's abilities, and an analysis of the classes being recommended. Each student must be considered in light of his or her specific instructional needs.

LABELS

In chapter 1 the concept of stigma was introduced. The importance of this idea is underscored in this chapter on impairments, disabilities, and handicaps. There is no accepted manner or precise definition for such terms. Not only are politicians and lay people at a loss, but also parents of handicapped students and citizens with disabilities themselves often use terms interchangeably or in ways connoting stigma. Additionally, social pressure causes us to label people. For example, many teachers and parents mistake the ability to use current parlance, initials, or technical words as a demonstration of expertise. In fact, these words are only global at best in terms of their descriptive ability. The amount of time saved in using labels is often outweighed by the potential of words to stereotype someone.

For example, in referring to a person as a "quad" rather than as "a person who uses a wheelchair" or "a person with special mobility needs," the image of a person existing within only a limited context comes to mind. On the other hand, functional descriptions do not imply present or future predictions of ability. Thus, describing someone as "needing assistance with reading" or "requiring more time to finish assignments" leaves a person to be a person and not just an "emr" (educable mentally retarded). By referring to a program as "vocational rehabilitation" instead of "vocational training/retraining" one may be literally adding insult to injury. Finally, there are the political pressures to contend with. Legislation often provides services only on the basis of a classification or labeling scheme. Thus, the one-dimensional view of an individual: the disability becomes the only image seen under the bureaucratic microscope. Disabilities receive services and people come attached to them.

The essence of any individual can be seen in his or her total of ability and inability. If this is kept in mind, then we can see that we are all handicapped in one area or another. That is, we all have a unique set of abilities and limits that help to set us apart as individuals. Many of us are able to disguise our weaknesses or limits and play up our strengths. Some have visible physical limits due to nature's random game of genetics or accident. Still others have learned inappropriate behaviors through a lack of proper training. For the latter groups it is not so easy to put on the guise of normalcy.

Unfortunately, this has led to their segregation within society. Yet we all have similar needs and aspirations to work, to live in the community, to enjoy leisure time. Ironically, it has been written in the popular press that 80 percent of the people alive today who live into old age will eventually experience an impairment that leads to a disability, further underscoring our similarities within a life cycle.

TERMINOLOGY

While disease is a concept that overlaps disability, the terms are not synonymous. First of all an *impairment* is a functional limit, loss, or restriction that may result from congenital defects, injury, disease, or environment. An impairment may or may not lead to a *disability*, or inability to perform a key life function. For example, a person with a hearing loss may not experience a disability in the auto shop because he or she can use visual indicators for motor performance (a strobe light, meter, or oscilloscope) and get enough information from the regular readings and demonstrations. For many persons the presence or degree of disability is a personal matter. For example, some people with hearing impairments enjoy going to dances while others stay away from such events.

When a disability impedes an individual's goal for travel, work, or normal living, he or she is said to have a *handicap*. This concept is based on the personal reactions of people in the surrounding environment, as well as the handicapped person's own view of the disability. For example, a wood shop teacher who refuses to take "those weird kids from the e m r room" becomes the primary handicap for students who are interested in taking the wood class. Also, Jean Edwards, an innovator in vocational education for the severly handicapped and a leader in handicap advocacy, has commented that we must not forget that

> . . . .Not all children who have disabilities are necessarily handicapped. It is important to remember that persons with disabilities are often gifted as well. Some prominent examples would include Helen Keller . . . Franklin D. Roosevelt . . . and rock star Stevie Wonder (Kimeldorf 1980).

If self-concept and identity are established through the major roles and relationships one has in life, then the need for normalizing educational experiences is further highlighted. The absence of meaningful work roles, leisure outlets, and regular classes deprives one of a source of social validation. Unemployed persons who were rejected from vocational rehabilitation agencies commented about their lives in this manner:

> "I live in a dead world".
> "I used to have a good disposition and helped others, now I am helpless . . . and I am bitter and grumpy."
> "Just sitting in a chair and looking like a human being."
> "I think they think I can't do anything".
>
> (The Urban Institute, 1975; p. 346)

Without regard to the specific disabilities, one might better view persons in functional terms. One application of this idea is to use functional terms that describe the degree to which an individual experiences a handicap. Persons with mild handicaps comprise the largest number of disabled Americans, up to 80 percent of disabled people in some instances. These students require very little adaptation of regular teaching techniques or placements. Mildly handicapped students generally do not want to be identified or singled out on the basis of their handicaps. Many students might feel reluctant to accept specialized and needed services in the presence of nonhandicapped peers. This is understandable in light of the fact that they may only experience a handicapping situation when in school where, for example, their reading ability has been diagnosed as a disability. Outside of school they bear no visually apparent handicap and consequent stigma. In fairness, it must be pointed out that many students with mild handicaps will not need special services within the technology curriculum.

At the other end of the spectrum are persons with severe or multiple handicaps. Functionally, they need educational, social, psychological, and medical services beyond the traditional regular and special education programs due to the intensity of their disabilities in order to maximize their potential for useful and meaningful participation in society or school.

Seen within this light, one can begin to reckon with the need for a continuum of services from pre-entry level classes where specialized programs and settings might be useful for a severely handicapped student, to entry-level and exploratory classes where a mild or moderately handicapped student might receive appropriate instruction. However, no student should be prejudged or placed solely on the basis of his or her degree of handicap. Each student needs the opportunity to receive appropriate instruction and the opportunity to demonstrate his or her ability. Again, quoting Jean Edwards:

> Given the appropriate education and training program, disabled persons should no longer be considered handicapped because with the right skills and opportunities they should appear more alike than different from the mainstream of society. It should be the goal of special education, I believe, to make it possible for everyone labeled as handicapped to acquire the tools necessary for blending and finding acceptance within our society at large (Kimeldorf, 1980).

This chapter presents information about the different types of impairments that often become disabilities or handicaps. These are grouped as sensory impairments (visual and hearing impairments), physical impairments, and learning and emotional impairments. The major focus is on the mild to moderately disabled person since this is where most "handicapped" students fall. Chapters 4 and 5 provide instructional techniques relevant to the technology curriculum. The last section of Chapter 4 presents instructional techniques for severely handicapped students.

SENSORY IMPAIRMENTS

Individuals with visual and hearing impairments have much in common, especially concerning the raw data they receive from their environments. For these people, improvement in the richness and variety of their sensory experiences is essential, since sensory losses often produce isolation and insulation from the mainstream of society. This is turn requires remediation in the form of methods, materials, equipment, and instructional settings that either (1) highlight the visual or auditory message (e.g., large print, hearing aids, etc.), or (2) circumvent the affected sensory channel (e.g., braille, sign language, tele-typewriters, etc.). Successful teaching in a life-centered career program depends on the resourcefulness of the instructor in providing alternative and modified learning modes that best capture what is seen, heard, felt, tasted, and smelled.

Visually Handicapped

Incidence

It is noteworthy to remember that the vast majority of visually handicapped individuals comprise the mild to moderate range of severity of handicapping condition. Thus, most will have some residual vision that can and must be utilized. Since no prosthetic sensory system can even remotely approximate the sensitivity and resiliency of our vision, it becomes even more essential that we train students to develop their vision (and hearing) to their fullest extent.

One erroneous notion held by many sighted persons concerning the visually handicapped is that they have little or no useful vision. In fact, the majority of visually limited people do have some residual vision, much of which can be trained for utilization in everyday life. Teachers, parents, and significant others can play an important role in helping these people use their remaining sight for improved functioning at home, in school, and in the community. Furthermore, it should be remembered that blindness is a problem of adults, and that less than 10 percent of the total blind population is of school age.

More than ever before, teachers are encountering visually handicapped students with additional disabilities. This can be any combination of deficits, such as a hearing impairment, motor problem, or emotional difficulty. How these disabilities interact will play a determining role in the educational approach to be taken. As more of these students are mainstreamed, it is essential that teachers receive help from the special teacher for program planning. Some of these students will be able to participate in the competitive marketplace and others will need more of a sheltered environment.

An individualized educational plan will be needed, and this will require close cooperation between regular and special class teachers. This cooperation will also be required when vocational training programs are to be developed.

Common Misconceptions

1. The visually handicapped have a better sense of hearing and touch than their sighted peers.
   Significant vision loss does not result in automatic improvement of any other sense. Training, practice, and continued use develop the other senses. Normally sighted people can be trained to develop these senses to the same extent.
2. All blind people read braille.
   There is no research that indicates a strong link between the amount of vision and mode of reading. Only a relatively small percentage of the visually handicapped population are braille readers. Sykes (1971) concluded in his study of visually handicapped teenagers that they performed as well with regular print as they did with large print, given optimum lighting conditions. Increasing evidence indicates that many legally blind individuals should be using the printed word.
3. Vision and reading are synonymous.
   While both processes are functions of the brain, reading also involves perceptual skills that (a) give meaning to what is seen, (b) relate it to past experiences, (c) store and retrieve the information, and (d) translate the above data into a decision-making system.
4. Eyeglasses (or contact lenses) strengthen one's eyes.
   Lenses do not alter the existing organic pathology, nor do they affect the progression of a disease process. Lenses refract or bend light so that a sharper, clearer image is produced on the retina of the human eye.

Terminology

For many years, teachers have noted that two people with the same visual acuity may have far different visual abilities—one using his or her vision, and the other using little or no vision at all. For this reason, Barraga (1964) coined the term visual efficiency to refer to how well a person uses his vision for the ordinary purposes of everyday living. The visual acuity reading found on an eye report from an ophthalmologist or optometrist *may not* be a valid indicator or predictor of how well that person will function in school or on the job.

Many overlapping and often confusing terms exist to describe the heterogeneous population called the visually handicapped. The first two, *legal blindness* and *partial sight,* are medical/legal definitions. The other two, *visual impairment* and *blindness,* are more educationally and behaviorally based. Legal blindness is

> a definition originally written into law to provide financial, legal, and social services to individuals with severe visual difficulties. People are considered legally blind when their visual acuity is 20/200 or less in the better eye after best correction, or they may have an acuity better than 20/200, but their visual fields are constricted to 20 degrees or less (*National Society for the Prevention of Blindness Fact Book,* 1966).

The designation 20/200 means that when tested at 20 feet, these individuals can see what a normally sighted person can distinguish at 200 feet from the eye chart. Best correction refers to the utilization of medicine, surgery, optical aids, or a combination of these types of treatment. The term 20/200 is *not* to be interpreted as a fractional equivalent of visual ability.

Partial sight or partial vision refers to a distance visual acuity better than 20/200 in the better eye, up to and including 20/70, after best correction. Visual impairment is a vision loss that leaves a learner with the ability to read the printed word.

Visual abilities are subject to many factors, and can fluctuate from time to time. Motivation, type, severity, and stability of the eye condition, available lighting, use of low vision aids, intelligence, appropriate instructional strategies, and additional disabilities are but some of the key factors that must be considered. Vision is by far our most complex sense, and because of this, it is often difficult to determine just how well an individual can or will see. All of the following factors must be evaluated in order to ascertain a person's level of eye functioning:

1. visual acuity—the sharpness of distance vision
2. visual field—the overall area seen by the eye(s) when in a fixed position
3. binocular vision—the ability of the eyes to function together to produce a single image
4. near or close-up vision—especially important for school work
5. night vision
6. color vision

Print size is designated regular (10 point) or large type (18 point). The print may or may not be supplemented with low vision aids (e.g., magnifying glasses, eyeglasses, closed circuit television, etc.). Regular type with low vision aids should be tried before selecting large type primarily because large type is more expensive, more difficult to produce and secure, and has fewer title selections than does regular print. The psychological implications of an alternative learning mode also may play a role here.

Blindness refers to such a limited visual ability that it would necessitate being a tactual reader. Braille, a system of raised dots, is used for this purpose.

The terms *visual handicap* and *visual limitation* are more inclusive and general, encompassing both visual impairment and blindness.

Potential Signs of Vision Problems

The following behaviors *do not necessarily* indicate a vision problem, but professional eye care and evaluation should be sought if any behavioral changes are noted with increasing regularity. It is essential that vision be tested annually if possible, especially when one already has a significant visual impairment. Equally important, all visually limited people should have their *hearing* checked regularly, since this sensory channel is crucial for their educational achievement. These potentially suspect behaviors include:

1. Recent aggressive, hostile, or withdrawn behavior.
2. Difficulty with reading or any work requiring attention to detail.
3. Difficulty seeing distant objects as clearly as before, such as blackboards. Problems in tracking moving objects across one's field of vision may be another important behavioral sign.
4. Tilting of the head or body when working or following directions. Check to see if one eye is preferred over the other.
5. Covering one eye when working.
6. Frequent complaints of headaches or double vision.
7. Coordination and/or balance problems. Poor motor abilities. Reports of increased difficulty in dealing with three-dimensional relationships.
8. Excessive blinking.
9. Holding work extremely close to the eyes. This may be accompanied by headaches, dizziness, rapid fatigue, and/or nausea.
10. Frequent complaints of eye pain, itching, burning, or crusting, or excessive eye rubbing.

Resources

Services for the visually handicapped are provided by numerous state, local, and national organizations. For a detailed listing of these groups, consult the *Directory of Agencies Serving the Visually Handicapped in the United States*, published by the American Foundation for the Blind.

Vocational services are provided by numerous agencies, including the state or regional rehabilitation center for the blind, the state commission for the blind, the state office of vocational rehabilitation, local sheltered and semi-sheltered workshops, and national advocacy groups, such as the American Council for the Blind or the National Federation of the Blind.

Low-vision services are available through many school systems, nearest medical school or school of optometry, the state commission for the blind and the American Foundation for the Blind.

Most (if not all) states provide *orientation and mobility services.* For more information contact the state commission for the blind, the state residential school for the blind, or the state department of education. In addition, many local school systems provide this service.

In addition to the above, specific services and organizations which provide them are listed below:

Service	Organizations
Education, rehabilitation, and research services including educational materials	American Association of Workers for the Blind; American Foundation for the Blind; American Printing House for the Blind; Association for the Education of the Visually Handicapped; National Society for the Prevention of Blindness; Hadley School for the Blind; Howe Press; Recordings for the Blind; Science for the Blind.
Library Services Braille, talking book machines, tape recordings, and periodicals	Library of Congress and the Regional Libraries for the Blind and Physically Handicapped.
Large Type	American Printing House for the Blind; National Braille Press; Stanwix House; check other major publishing houses. Large print typewriters are available from IBM, Smith-Corona, and other major producers.
Periodicals - Braille	American Printing House for the Blind; Clovernook Printing House for the Blind; National Braille Press.
Periodicals - Tape Recordings	Science for the Blind; Hadley School for the Blind.
Periodicals - Talking Book Machines	American Printing House for the Blind; American Foundation for the Blind; National Braille Press.

Integration

While some specialized approaches are directly geared toward difficulties arising from vision and hearing impairments, many of the instructional techniques and strategies applicable to regular students are likewise applicable to exceptional students. It is crucial, therefore, that teachers focus on the *similarities* these learners share with *all* students. Often they will find that most of their instructional programs can be readily adapted for use by handicapped learners, usually without significant modifications. Where modifications are necessary, or where highly specialized procedures are required, help can be readily received by consulting the resource room or itinerant teacher charged with coordinating services to handicapped students.

There is no such thing as a "blind personality" or "blind adjustment problems." The same qualitative problems encountered by sighted people are also encountered by the blind. Blind individuals are not necessarily more unhappy, depressed, or more anxious than their sighted peers. They are not more prophetic nor musically capable, but show the same spectrum of abilities shown by any other group of individuals. To understand this group best, one should look at how society reacts to minorities in general—the stereotypes, the attitudes, and the expectations that have evolved over many years.

Classroom Instructional Strategies

The sense of vision is the primary channel of learning for humans. When vision is reduced or absent, the remaining senses must be utilized to compensate for this reduction in environmental stimulation. The senses of touch and hearing, together with improved visual efficiency (if possible), must be the primary means of compensation. Hearing enables people to communicate verbally and maintain social contact. It is most important for visually handicapped learners to develop their listening skills, since much information must be presented in recorded form. Touch provides clues for indicating size, shape, and positions of objects. With this sense, however, reality must be constructed in a piecemeal fashion. Tactual units must be put together to perceive wholes. While neither sense is as efficient as vision, they nevertheless provide important substitute modes of learning when vision is compromised.

While specialized techniques, materials, and strategies may be employed, the same basic information intended for the normally sighted student can also reach the visually handicapped student. The following suggestions should be kept in mind:

1. Control over room lighting is crucial. A number of eye conditions result in photophobia or extreme sensitivity to light. Shades or blinds should be used to regulate the amount of light entering the room. Try not to obstruct the available light in the room. Do not seat students in a direct easterly direction in the morning, and above all, eliminate glare whenever possible. Proper lighting will help maximize residual vision, and should be used in conjunction with low vision aids.
2. Firsthand experiences with the environment are essential. Complete dependence on verbal explanations and models is not desirable. This should occur only when actual contact with the real object or situation is impossible or unfeasible. This probably will require closer supervision and more time in the initial stages of instruction, but in the long run it will pay dividends. All students should *learn by doing*. Tactual learning will require more time than visual learning, since the func-

tionally blind student must put pieces together to get whole pictures.

3. Provisions must be made for bulky equipment. Visually handicapped students will be transporting large, often unwieldy equipment to class. Much of this will require lots of storage space. Typewriter-like machines to produce braille (braillewriters), the Optacon (a tactile reading device), tape recorders, etc., not only necessitate storage space, but also a number of easily accessible electrical outlets. With younger students, safety caps placed over outlets will serve as an accident-preventing measure. When equipment is to be moved or classroom arrangement altered, always notify the visually handicapped student beforehand.
4. Audiovisual materials: When films, slides, and other print media are used, the blind student will be at a disadvantage. If possible, a sighted reader for the class will help the blind person gain access to the same information given to the rest of the class. Try to select audiovisual materials that have an excellent audio component.
5. Low-vision aids: Students with certain eye conditions do well with magnified materials, and a number of low-vision aids would be appropriate for these students. Often, aids are relatively inexpensive and easy to locate. They include:

   - Eyeglasses and contact lenses. Bifocals and jeweler's loupes may be helpful for near vision.
   - Hand-held magnifying glasses or mounted magnifiers with illumination.
   - Closed circuit television (CCTV), such as the Apollo Laser and Visualtek.
   - Projection magnification, such as through the use of overhead projectors.
   - Telescopes, either hand held or mounted in eyeglasses for distance viewing.

   Low-vision aids, like hearing aids, will not automatically produce gains in the classroom. Using them requires practice, motivation, adjustment, and proper maintenance. It is important to remember that as magnification increases, the field of view is correspondingly decreased. This will necessitate adjustments, primarily increasing the time taken to master overall visual dimensions and organization. The use of illuminated reading stands is also recommended. For further information, the reader is directed to the educational and low-vision implications sections in *The Low Vision Patient* by E. Faye, and *Visual Impairment in the Schools* by R. Harley and G. Lawrence.
6. Braille materials: For braille readers, try to select texts that are available in both type and braille on a commercial basis. When the braille version is unavailable, local civic or religious groups may provide voluntary braille transcribing services. It is wise

to plan ahead at least six months for securing a braille version of your class text, especially if it must be transcribed by volunteers. Transcribing classes may be offered locally or by mail from the Library of Congress in Washington, D.C. Successful completion of the transcribing class is dependent upon performance on a proficiency examination.

7. Resources: The largest supplier of educational materials for the visually handicapped is the American Printing House for the Blind in Louisville, Kentucky. Students are entitled to materials based on a quota account established by the federal government. Check with the special teacher to confirm that your students are entitled to receive this service. Where special materials are needed (for blueprint production, charts, diagrams, writing materials, etc.), consult the American Printing House catalogs.
8. Coactive instruction: For visually handicapped students, especially the totally blind, it is strongly recommended that the instructor perform the task *with the student*. The teacher should have the student's hands in his, moving together as a single unit. This is particularly useful at the outset of a new instructional sequence. This hand-over-hand approach should be supplemented with verbal explanations, and do not assume that a blind student has a concept well mastered by verbal demonstration alone. Assess competence by actual, on-going performance abilities. During instruction, always use students' names, and describe settings so that the blind student will have a proper orientation to classroom activities.

Technology Lab Instructional Strategies

Specific suggestions will depend on the disabilities of the individual students. In general, however, safe operation and use of equipment, detailed plans (blueprints) and instructions, and measuring devices adapted for use by handicapped persons are items to be considered in all cases.

1. Safety procedures: Fire drill procedures should be carefully explained. With practice, the visually handicapped student, with or without assistance, should be able to reach safety in the event of a fire. Teachers may wish to implement a "buddy" system for all students. For partially sighted pupils, large decals applied to window or door surfaces may help prevent serious accidents. Glass doors and sliding windows should be completely open or shut. Make sure to orient the students to the overall terrain of the school, emphasizing unsafe areas. Insist that safety equipment be worn when operating potentially dangerous machinery, and establish procedures for reporting accidents.
2. Orientation and mobility: Orientation is defined as the ability to localize oneself in relationship to the environment, while mobil-

ity deals with movement from one place to another. Rooms should be organized so that there is a well-defined traffic pattern. Teachers can learn some basic orientation and mobility techniques from specialists who visit the school on a regularly scheduled basis to work with visually handicapped students.

Aisles should be made wider than usual, especially if there are other students with movement problems. Give students plenty of time to explore the room, and help them develop a mental picture of the physical arrangement. This can be done by constructing tactual or relief maps, and most importantly, by providing first-hand tactual and auditory experiences.

If furniture or equipment is to be moved, make sure that the student is kept informed. Keep wastebaskets under desks, and check to see that aisles are kept clear at all times. Whenever possible, encourage students to be independent travelers.

3. Power machinery: For any type of exploration to be successful, a systematic search pattern must be adopted. For power equipment, this generally involves a two stage process:

   a. General orientation to the equipment: This includes exploration of the front, back, sides, top, and bottom of the device. It also involves correct body positioning with respect to the front of the equipment, and a thorough knowledge of the spatial relationships of the controls for operation of the device. When this stage is successfully learned, the student should have a clear mental picture of the overall structure of the equipment, as well as the individual controls.

   b. Orientation to the functioning and moving surfaces: This involves operation of the power switch(es); correct positioning of the workpiece; knowledge of the path of the cutting edge or blades; and, knowledge of safety procedures.

One way to facilitate exploration (and identification) in both stages is to visualize the overall equipment or individual surfaces as a clock, with parts at certain hour designations (e.g., on/off switch at three o'clock, cutting path from twelve to six o'clock, etc.). When working with the student, place his/her hands in yours, and together move through each step until the task is done. This should be repeated until the instructor is confident that the student can perform the task independently. All dials and switches should have easy-to-read print or braille labels.

Cutting edges or blades should have adjustable, protective guards allowing proper movement of the workpiece. Provide visual, auditory, and tactual cues that facilitate proper use of the equipment. For example, check for distinctive features of the workpiece for correct positioning, or sound of the blade to indicate proper functioning.

Safety procedures should be closely followed. Again, use visual, auditory, and tactual cues to indicate machinery malfunction. Students should have a basic knowledge of (1) quick access to emergency shut-off, (2) a safe route of movement from the machinery, and (3) first aid.

2. Modifying plans and blueprints: Transforming plans and blueprints into three-dimensional tactual models is often most helpful. Perhaps the student can take dimensions of a full-size model and transfer them with gauges or calipers when blueprints would not be useful. Keep in mind that tactual exploration is slow, and models should be simple and uncluttered with extraneous features. A proper scale of distance should be followed when using scale models, and an easy-to-read brailled key or legend must be provided. Use materials that will withstand rough handling. Tactual plansheets can be constructed using wire, styrofoam, fabric, tape, etc., and copies can be produced using the American Thermoform Duplicator and special paper called Braillon (see appendix D).
3. Measuring and calculating: Examples of devices available include:

   - Braille rulers, compasses, balances, raised-line drawing kits, drafting supplies, embossed graph paper, etc., from the American Printing House for the Blind (APH), 1839 Frankfort Avenue, Louisville, KY 40206.
   - Protractor and sawguide combination, modified Starrett Micrometer, drill guides, wood marking gauges, etc., from the American Foundation for the Blind (AFB), Aids and Appliances Catalog, 15 West 16th St., New York, NY 10011.
   - Laboratory, darkroom, and clinical thermometers, light probes, and meter readers with audible signals and/or braille read-outs are available from Science for the Blind.
   - The "talking calculator", available on quota account from the American Printing House for the Blind, Louisville, Kentucky, or directly from Telesensory Systems, Inc., Palo Alto, California. Performing a number of mathematical functions, this calculator has a synthesized speech output. Its importance lies not only for computation, but it opens career opportunities for the blind person in the sciences and higher mathematics.

Hearing Impaired

Terminology

Whereas a legal definition exists for blindness, no such definition of deafness is available. The term *hearing impaired* is a general one, and may indicate any degree of hearing loss—mild, moderate, severe, or profound. The term *hard of hearing* refers to one who uses his or her hearing to develop speech and language (with or without amplification). On the other hand, the *deaf* usually will not be able to use their hearing for this purpose, even with amplification.

Hearing enables people to explore and participate in their environments, while providing a key vehicle for social interaction. Without appropriate intervention, hearing impairments can have a substantial negative impact not only on the development of communication skills, but in the emotional and cognitive areas of human growth and development as well. For example, understanding and using abstract concepts is often difficult for deaf students (e.g., technology forecasting), as is the use of idiomatic expressions (e.g., "horseplay" or "Keep the gauge 'true' at all times"). This has extremely important implications in the areas of job satisfaction, leisure activities, and the social impact of our technological advances.

Common Misconceptions

1. Hearing impaired people have a keener sense of vision than normally hearing peers.
   Hearing loss does not automatically result in improved vision. Individuals with hearing problems must rely on visual clues to develop properly and keep in constant touch with their environment.
2. Deaf people are also mentally retarded.
   Unfortunately, the term "deaf and dumb" (implying a lack of intelligence) is still used today. Hearing impaired persons, especially those with poor speech or those who communicate manually, may be incorrectly labeled mentally deficient. There is no clear-cut causal relationship between hearing loss and depressed cognitive abilities. Environmental retardation due to limited opportunity is a more common and insidious problem.
3. Speech reading or lipreading can fully compensate for a hearing loss.
   At best, distinguishing sounds by watching the lips of the speaker can provide understanding of only about 25 to 30 percent of what is spoken. One reason for this is that many of the words in the English language are homophonous, i.e., while they sound differently, they look similar on the speaker's lips (e.g., male, bale, pale). Furthermore, factors such as lighting, distance from speaker, poor speech, reduced lip movement, fatigue, etc., all detract from higher intelligibility rates.
4. Hearing aids quickly improve hearing ability and communication skills.
   For this to occur, aids must be worn in the classroom, at home, and in the community—in short, as often as possible. The use of hearing aids should be supplemented with speechreading and auditory training to help improve listening skills and for better speech discrimination. This will take practice, training, opportunity, and most importantly, *time*.

Potential Signs of Hearing Problems

The following behaviors may or may not indicate a hearing loss. If you suspect a decrease in hearing acuity, consult the special education teacher. Referrals can then be made to an audiologist or otolaryngologist for a comprehensive hearing evaluation. These suspect behaviors include:

1. Increased difficulty in hearing distant or close-up sounds in recent times. Note hearing changes for speech in the conversational range.
2. Lack or change in response to loud sounds.
3. Inability to screen out extraneous sounds.
4. Complaints of "ringing in the ears" or pain in the ears.
5. Problems in monitoring one's voice and speech—may be exceptionally loud or soft.
6. Tendency to become more dependent on vision in learning situations.
7. Aggressive, "acting out" behavior, or withdrawn, passive behavior.
8. Less coordination of visual and auditory clues (looking in direction of a sound).
9. Speech and language regression.
10. Less interest in social interaction.

How a hearing impairment affects a learner's educational progress is largely determined by the following factors.

1. The nature of the hearing defect. Greater problems will occur when the hearing loss is in the normal speech range, where most of our conversation takes place. Losses in the higher or lower frequencies pose less of a functional problem.
2. The degree of the hearing defect. Profound hearing losses usually preclude help through hearing aids, whereas milder losses can be helped significantly through amplification.
3. The onset of the hearing loss. Generally, the earlier in life the hearing loss occurs, especially in the first year or two, the greater the educational difficulties. Occassionally, there will be an exception to this.
4. The level of a person's intelligence. Higher intellectual levels result in a greater probability of academic success.
5. The quality and quantity of stimulation given to the student. Early sensory stimulation based on sound educational principles will maximize opportunities for self-fulfillment and success in learning.

Resources

The following list includes only some of the many agencies providing services to the hearing impaired. For a more detailed listing, consult the special education teacher or vocational rehabilitation counselor.

Service Provided	Agency
Resources for teaching speech reading and speech; publishes *Volta Review.*	Alexander Graham Bell Association for the Deaf, 3417 Volta Place, Washington, DC 20007
Education programs for deaf and deaf-blind children and their parents; maintains diagnostic and research center on deafness.	John Tracy Clinic, 806 W. Adams Boulevard, Los Angeles, CA 90007
Organization predominantly of the deaf to promote programs and services to hearing impaired individuals and parents.	National Association of the Deaf, 814 Thayer Avenue, Silver Springs, MD 20910
Vocational and technical programs of higher education; information resources for those interested in working with the hearing impaired.	National Technical Institute for the Deaf, One Lomb Memorial Drive, Rochester, NY 14623
Rehabilitation services; referral source; publishes *Journal of Rehabilitation of the Deaf* and *Deafness Annual.*	Professional Rehabilitation Workers with the Adult Deaf, 814 Thayer Avenue, Silver Springs, MD 20910
Provides a wide range of services—educational, social, and public services, community action, etc.	Gallaudet College, 7th and Florida Avenue, N.E., Washington, DC 20002
Free captioned films, as well as other media	Media Services and Captioned Films Branch, Bureau of Education for the Handicapped, 7th and D Streets, Washington, DC 20202
Rehabilitaion and Vocational Services	Office of Deafness and Communication Disorders, Department of Health, Education, and Welfare, Washington, DC. Also contact your state office of vocational rehabilitation, Goodwill Industries, and local teacher of the hearing impaired.

Modified from L. Katz et al, *The Deaf Child in the Public Schools,* Interstate Printers, 1974.

Integration

Hearing impaired individuals represent a heterogeneous population, with a wide range of abilities, just like those with normal hearing. Be careful not to stereotype this group on the basis of limited interaction opportunities.

Reduced communication skills, limited accessibility to services, unfortunate stereotypes, and intolerant attitudes all contribute to feelings of isolation and insulation. It is easily understandable how retreat into a "deaf world" could thus occur. Clannishness is more a result of the availability of programs and services, as well as coping skills, than deafness per se.

Classmates of hearing impaired students should be acquainted with the nature of hearing losses, and how potential problems can be overcome. In-service education for hearing students and other teachers can have far-reaching, positive results. It is crucial, especially at the outset of the academic year, that the hearing impaired pupil develop and maintain friendships with classmates. Careful monitoring of the classroom/social environment may be an important preventive measure for avoiding social/emotional problems, as well as ensuing academic difficulties.

Classroom Instructional Strategies and Aids

Some hearing impaired individuals have problems in developing language and reading skills. The following suggestions are provided to help minimize these problems while keeping the student in closer touch with the overall classroom environment.

1. Have the student sit closer to the teacher, where residual hearing can be used to its utmost. Make provisions so that the pupil has an unobstructed view of the teacher and any key instructional material or equipment.
2. Check with the special teacher to determine if hearing aids, speech reading, or any other special approach has been tried. Learn how to check hearing aids, and help students maintain them properly. If prescribed, insist on their use, and motivate students to wear them.
3. Since hearing aids generally amplify all sounds, try to reduce extraneous noise, especially during instructional periods.
4. When using audiovisual materials, select *captioned films* whenever possible. These films are supplemented with printed subtitles. Pupils will enjoy and learn from them, and the special teacher can help in ordering these films.
5. Try to augment verbal explanations with pictures, charts, diagrams, and other graphic materials. Direct, firsthand experiences are preferred over the use of models and replicas of the real thing.
6. Key words, central themes, safety rules, and idiomatic expressions (e.g. "hands-on, buddy system", etc.) should be written out. Do not be reluctant to use natural gestures and signs if you feel that they would facilitate learning.
7. Since hearing impaired students depend to a large extent on their vision, maintain proper lighting conditions. Light-colored walls with a matted finish will help reflect existing light, while minimizing glare.
8. If deaf students need to take notes, but depend on visual cues from the instructor, ask a classmate to take a second set of notes using carbon paper. This would be applicable for the visually handicapped student as well.
9. When speaking to students with hearing difficulties, don't exaggerate words or gestures; present material at a normal rate, and use full sentences. Don't hesitate to repeat central themes or key thoughts. Require that students speak in full sentences.
10. It would be helpful, especially when presenting complex material, to prepare a brief set of notes for additional study. When using dittos, make sure that they are clear and easy to read. Try to summarize information for the student using charts, diagrams, pictures, etc.

11. Help the student become knowledgeable of the physical surroundings of the room. If any special equipment requires close supervision, provide additional time for teaching proper operation and maintenance.

Hearing Aids

The purpose of the hearing aid is to increase the intelligibility of speech. This does not automatically occur once the aid is fitted. Just as with the low vision aid, it takes motivation, practice, and time to use effectively. It will not help all types of hearing losses, nor can it totally compensate for a hearing loss. Whenever possible, hearing aids should be supplemented with auditory training and lipreading (speech reading).

Regular classroom/shop teachers should work closely with the special education teacher to help the student use the hearing aid to its maximum potential. This includes:

1. Monitoring the student's use of the aid. Proper insertion of the mold, maintenance and replacement of batteries and cords, and periodic checks for distortion and mechanical problems.
2. Carefully recorded observations of behaviors that may signal changes in hearing acuity.
3. Explaining to other students in the class how the aid works, its care, and importance (this can be done by both the teacher and student).
4. Reminders that the aid is not a toy, but a most useful device for improved function. To gain the most from the aid, *it must be consistently used.*

Speech reading

This is another important tool for effective communication. Good lipreaders do not catch every word of the speaker (less than one-third of the speech sounds are clearly visible), but depend on synthesizing key words and elements of the spoken message. It serves as a most helpful companion to a hearing aid.

To help those with a hearing loss lip-read more effectively, the following considerations should be taken into account:

1. Speak as normally as possible. Don't exaggerate mouth and lip movement, as well as gestures.
2. Beyond eight to ten feet, lipreading is extremely difficult.
3. Lighting is critical. Minimize glare, and see that the light falls on the speaker's face.
4. Chewing something while talking makes it more difficult for the student to read your message.
5. Lipreading straight ahead is easier than at a steep angle.
6. Lipstick tends to frame the lips, making them easier to read. Certain styles of beards and moustaches may hinder lipreading.

7. Flickering lights and sunglasses are distracting.
8. Even under optimum conditions, speech reading is an intensive and tiring process. Keep lectures and presentations to reasonable lengths.

Oral vs. Manual Approaches

For many people with limited hearing, the use of residual auditory abilities for speech and language development is vital. For others, a manual approach is necessary. This may include the use of the manual alphabet (finger spelling) and/or sign language. For still others, a combination of the two approaches may be most appropriate. Examples include Total Communications and the Rochester Method. Keep an open mind about which approach works best for each person. Controversy over the best method has raged for many years, and only individual learner needs and abilities will dictate the most useful selection. It would be most beneficial for the regular teacher (or any interested instructor) to have a basic understanding of finger spelling and sign language. A short period of in-service instruction from the special teacher could go a long way in providing these skills.

Figure 3–1. *Assistant professor Dominic Bozzelli (seated) shadow teaches an engineering class with assistant professor Kevin Foley. (Courtesy of National Technical Institute for the Deaf, Rochester, New York.)*

Approach	Emphasis/Philosophy
Oral	Person must function in a world of sound. Speech and language training are supplemented with auditory training, speech reading, and amplification.
Manual (Amesian: American Sign Language; Finger Spelling: Manual Alphabet; Cued Speech: Speech Reading plus Hand Signals)	Will use gestures, signs, cues, etc., for communication. Depending on system, syntax may be altered (i.e., tenses omitted, word order changed, etc.)
Combination (Total Communication: Speech plus Finger Spelling and Sign Language; Rochester Method: Speech plus Finger Spelling)	Several systems used simultaneously to capitalize on residual hearing as well as the strengths of the manual approach.

Technology Lab Instructional Strategies

1. For first drill procedures, a light placed over the fire alarm (operating simultaneously with it) in full view is helpful. Also, insist on the use of safety lenses when operating any potentially dangerous equipment.
2. Teletype machines (TTY) are typewriter-like devices connected to one another by telephone lines. A person simply types a message on his/her machine and it is carried to someone else with a TTY that the sender has signaled. A national TTY directory is maintained and updated periodically by Teletypewriters for the Deaf, Inc., and the use of this tool is growing rapidly among the deaf. It has far-reaching educational and vocational potential, especially in computer-assisted instruction.
3. Rules for the operation of equipment should be easily seen and clearly written. Simple pictures or diagrams will be helpful.
4. Some blueprints and plans may require models or replicas for instructional purposes, especially where extensive descriptions and complex terms are used. In these instances, simplifying work requirements by paraphrasing or summarizing directions and explanations should prove helpful.
5. Like all students, the hearing impaired learn best by direct, firsthand experience. Do not presuppose that the student has mastered a skill by verbal or manual communication responses only. Carefully observe the student's performance, and give him/her the necessary feedback.
6. In certain settings, such as laboratory work, use of the buddy system should be considered. This may be especially valuable in situations where audiovisual materials are lacking or where complex verbal demands are made on the student.
7. Try to select curriculum materials that use excellent pictures or diagrams to explain essential features of the lessons. Some of these materials are made expressly for the hearing impaired (e.g., *New Jersey Vocational-Technical Curriculum* available from Rutgers University, New Brunswick).

8. Students should be counseled regarding postsecondary education and training options. Although many institutions provide interpreters, two institutions are specifically geared to the needs of the deaf and hearing impaired. Gallaudet College in Washington, D.C., is a liberal arts college funded by the federal government. It offers a wide variety of educational, social service, and research programs to promote the welfare of those with hearing problems. The National Institute for the Deaf, located on the campus of the Rochester Institute of Technology in New York State, provides programs in the technical fields, including engineering, graphic art, and commercial photography.

PHYSICAL IMPAIRMENTS

It is the goal of every educational program to develop those skills that will permit a student to function independently within our society. Every disabled student should also be permitted every possible opportunity for similar growth and development. The opportunity for a disabled student to participate in a technology education curriculum is an important step in this direction.

Misconceptions

One misconception that most people have about the physically disabled needs to be discussed. Our abilities to effectively communicate are measures by which people unconsciously judge our intelligence. In many cases, a person's disability may affect and limit his communication skills. If those skills are severely impaired, a physically disabled person becomes isolated and limited in his interaction with the environment. These limitations can easily convey the impression that a physically disabled person is also mentally retarded, which is often not the case. A person can also appear mentally retarded because his physical limitations result in a reduced ability to interact with his environment. Teachers should not assume that a person may have congnitive impairments or delays because of communication impairments. Otherwise, inappropriate teaching materials might be selected. A teacher needs to view the physically disabled as having the ability to learn everything that is brought before them, and this belief should be strictly held to unless proven otherwise.

Terminology

The following section primarily deals with descriptions of some of the major physical disabilities that can be encountered. These descriptions are not designed to provide the reader with a total understanding of each disability. They are designed to give a general understanding of the specific condition and to provide some practical knowledge that will be useful in the classroom.

Cerebral Palsy

To quote *Physically Handicapped Children: A Medical Atlas for Teachers:* "By definition and common usage, cerebral palsy is a nonprogressive disorder of movement or posture beginning in childhood due to a malfunction of the brain" (Bleck and Nagel 1975).

What should be noted in this definition is that cerebral palsy (C.P.) is referred to as a *disorder* rather than a disease. This distinction is important because too often C.P. is grouped with such illnesses as polio, meningitis, and the common cold. These illnesses are caused by viruses or bacteria while C.P. results from damage inflicted on the brain or the spinal cord.

This damage can occur while a child is developing (prenatal), during the actual birth process (perinatal), and within two years after birth (postnatal). C.P. can result from a tumor in the brain cavity; a metabolic disorder, such as diabetes; anoxia, which refers to a lack of oxygen being delivered to the brain shortly after the birth; a traumatic injury to the head; or a prolonged period of labor. While encephalitis is medically defined as a disease, a child can develop C.P. as a result of it.

Cerebral palsy is classified by five categories based upon the differing movements that the person displays. It is very rare to find a person diagnosed as having C.P. and only displaying one type of movement. The majority of students will show two or more differing movements; however, a person will be diagnosed according to the most dominant movement first (e.g., spastic ataxic). Such a classification indicates that movement is dominated by spastic movements with the additional presence of some ataxic movements.

Not only will a child's disability be diagnosed according to the most dominant movement, it will also be characterized by the number of limbs affected, e.g., *spastic ataxic triplegic.* A list of the more commonly used terms follows:

Monoplegic: One limb is affected.
Hemiplegic: An arm and a leg on the same side of the body are affected.
Paraplegic: Only the legs are affected.
Diplegic: The major involvement is with the legs and the arms are affected less.
Triplegic: Three limbs are affected, usually both legs and one arm.
Quadriplegic: All four limbs are affected.

Cerebral palsy not only affects a person's limbs, it can also affect other functions of the body. It can affect a person's eye muscles or the muscles that control the mouth. An accompanying disability can involve a person's vision or speech. In some cases a person's hearing is involved as well. Depending on the severity of a child's physical disability and accompanying disabilities, his or her ability to effectively interact with the surrounding environment could be seriously limited. If this limited interaction is permitted to continue for an extended period of time, mental retardation

Movements	Characteristics	Useful Information
Spasticity	1. Muscles display an increased amount of tension. 2. Relaxation is difficult. 3. Movements appear stiff and awkward.	1. The most dominant form of C.P. 2. A high incidence of mental retardation accompanies spasticity. 3. Surgery is often used to relieve severe cases of muscle tension.
Athetosis	1. Movements appear involuntary and purposeless. 2. Muscles may appear tense, but will relax after repeated movement. 3. Muscles that are not directly involved in a desired movement will conflict with a goal-directed movement, e.g., the left arm may cause the body to be mispositioned as the right hand attempts to grab an object.	1. Speech impairments can also be present.
Rigidity	1. Muscles move with great difficulty due to severe tension. 2. All four limbs are usually affected.	1. Most severe form of C.P.
Ataxia	1. Loss of balance. 2. Movements are uncoordinated. 3. Child lacks spatial awareness between the body and immediate surroundings.	1. Only way of maintaining balance is to walk with outstretched arms and legs.
Tremor	1. Shakiness in an arm or a leg. 2. Shakiness may appear as the limb is in motion.	1. It is rare that a limb would be continuously in motion.

could also develop. When a person has cerebral palsy and an additional disability, then the person comes under the category of services designated for the multi-handicapped.

The degree of severity for each type of cerebral palsy varies with the individual. A person with a mild involvement is one whose disability does not require any corrective devices, such as crutches, braces, or wheelchairs. Some therapy may be required, but it would only occur for a short period. This person would be fairly independent and require very little assistance.

A person with moderate involvement usually requires the use of a corrective device and extensive therapy may be required. The ultimate goal of the therapy is to develop the person's capabilities so that independent activity can be achieved. Most of the therapy or surgery occurs during a child's primary years, and by the time a person reaches the secondary level, such corrective procedures would normally not be needed. A person with a moderate degree of cerebral palsy could eventually lead an independent life and function in a technology class with minimal assistance.

A person with a severe case of cerebral palsy is generally affected in most of the limbs and an additional disability is present. The person probably requires some surgery in order to relieve muscle tension or develop better muscle control. Depending on the severity of the disability, such a person could still benefit from a shop experience. The person may require one-to-one instruction in order to accomplish some of the class exercises. For some, a shop experience could be limited to just being exposed to the workings of different technologies. Everyone can benefit from becoming technically literate to the degree he or she can participate. An awareness of technical processes may serve a student later in careers related to computer programming, engineering, sales, or other activities related to consumership. A person's self-concept is enhanced when he or she can demonstrate a knowledge of materials and processes to others. Any attempt to determine which is the least restrictive environment should be made jointly by the student and/or parent, special education teacher, and technology instructor.

Rehabilitation varies with each individual and the severity of the disability. The goal of any rehabilitation program is to develop the greatest degree of independence. Occupational and physical therapy are the main components. These therapies concentrate on developing an individual's use of his or her arms and legs. Braces, crutches, and muscle surgery may also be required to develop that usage.

The long-term outlook for a person with cerebral palsy varies according to the severity of the disability. Some individuals are capable of adjusting to the demands of everyday life without any difficulty. Others can achieve the same adjustment with some modifications, such as the usage of a corrective device or the assistance of an attendant in their homes during their adult lives. There are other individuals whose disability is so severe as to require constant supervision, and their ability to functionally interact within society is greatly curtailed. For those individuals the eventual goal may be to live in a group facility and work in a sheltered environment. But no matter how severe the disability, it is important to recall that every person has the potential to accomplish many fulfilling activities during his or her lifetime.

Epilepsy

Physical disabilities are usually considered observable and very apparent. In most cases this assumption is accurate, but in one major instance it is not. This major exception is *epilepsy*. Although it is not commonly viewed as a physical disability, its effect is as profound as a severe case of cerebral palsy and that is why it is included in this chapter.

Epilepsy is one of the most misunderstood, and often feared, disorders that can affect a person. The number of misconceptions, fears, and injustices associated with epilepsy are too numerous to list. This information is restricted to a discussion of the basic facts concerning epilepsy and of what should be done if a seizure occurs.

Epilepsy, like cerebral palsy, is a disorder of the brain and is associated with convulsive occurrences. What causes epilepsy is still unknown. Specialists believe that the brain is damaged as a result of internal bleeding, inflammation of the brain (encephalitis), trauma, brain tumors, or a metabolic disorder. All of these conditions may contribute to the formation of a lesion that leads to the appearance of epileptic seizures. The lesion does not cause the seizure for it is an area of dead brain cells. The cells surrounding the lesion cause a seizure for they are damaged and function improperly.

One way of understanding a seizure is with this illustration. The brain produces electrical impulses in its daily operation of controlling the body's activities. When a lesion is formed, the electrical impulses from the surrounding cells are disrupted. At times these disrupted impulses become overloaded and a seizure occurs.

When a person thinks of a seizure, he usually will think of the most severe form of convulsive activity stereotyped as a person falling to the floor, head banging severely, and jerking movements of the body. This form of a seizure is known as a *grand mal* seizure. Along with the previously described characteristics, a person may lose consciousness, the body will stiffen, and occasionally a loss of bladder control may occur. If a grand mal seizure occurs, here are a number of steps that can be taken to insure the person's safety:

1. Chances are very slim that you will be able to catch a person as he or she falls.
2. Do not attempt to stop the seizure once it begins. Permit it to "run its course." It should last for no more than one minute. If it continues for more than three minutes, get medical assistance as quickly as possible.
3. Remain calm.

4. Place a pillow, a rolled up rag or coat, or a foam pad under the head to prevent injury if any movements occur.
5. Remove any freestanding objects around the person. They may injure him or her if the body involuntarily moves.
6. Do not worry about the tongue. The jaw will lock shut during a seizure and nothing can be done to open it. Placing a pencil in the mouth only increases the possibility of injuring the person for the jaw will shut and splinter the pencil.
7. After a seizure if there appears to be difficulty in breathing, check the mouth. The tongue may be blocking the breathing passage and should be placed in its proper position. This repositioning can be simply done by reaching for the tongue and pulling it forward.
8. Do not have other students crowd around the stricken student. It is upsetting to revive and find a dozen faces staring at you.
9. Have a friend or yourself near the student to comfort and reassure him/her when he/she revives.
10. When he or she revives, allow the student to rest for a few minutes and keep him/her warm with a blanket or coat.
11. The student may be disoriented for awhile. He or she needs to be told what has happened and reassured.
12. A grand mal seizure is a very tiring experience to endure and the student may not be able to work as hard as before having the seizure.
13. Inform the school nurse as soon as possible. Write down the events prior to and during the convulsion. This information may have medical relevance later.

When a seizure occurs, events happen very quickly. If you have never witnessed a seizure, you will probably get involved observing it and forget the necessary safety steps. A recommendation would be to have a class discussion about epilepsy and seizures (in which the Epilepsy Foundation can assist). A role-playing exercise may be of benefit to the other students in the class.

Another type of seizure is known as *petit mal*. A petit mal seizure appears in a variety of forms. It can be either a momentary twitch on the face, a blank expression, a stiffening or jerking of an arm or finger, or a possible loss of consciousness. Unlike a grand mal seizure, a petit mal seizure only lasts a few seconds. In most instances it is over before anyone is aware of it. The student will be momentarily disoriented after the seizure, but within a few minutes he or she will readjust to the environment and continue with the normal routine. Because of their quickness there is very little that can be done except to reassure the student once the seizure has ended. Psychomotor seizures are a third type best described by talking to a professional since they vary so much.

It has been emphasized that there is nothing a person can do in order to prevent a seizure from occurring, but there are particular things in a student's environment that should be avoided for they can unnecessarily cause a seizure. The lights on-and-off pattern disrupts the brain's electrical impulses. Overexertion, especially combined with becoming overheated, can contribute to the beginning of a seizure. These situations are very simple to avoid and require no major disruption of a technology classroom program.

Some people can anticipate a seizure and take some appropriate response. What these people are experiencing is known as an *aura*. To describe an aura is difficult, and the only adequate approximation is that the person suddenly has an unusual feeling. The aura occurs only a few minutes before the seizure begins and allows the person to briefly prepare himself or herself. This preparation consists of simply sitting in a chair or lying down and asking for a person to be nearby. Beyond this point, nothing else can be done.

Medication is the best means for controlling the frequency of seizures. The most frequently used drugs are phenobarbital and mephobarbital. These medications do not completely halt the occurrence of seizures, but they dramatically reduce their frequency. Students are constantly being monitored by physicians to determine the effectiveness of the dosage. Reporting the occurrence of a seizure to a school nurse is important for it is one way of monitoring the medication's effectiveness.

It should be expected that a teacher in a life-centered educational program could become apprehensive about the placement of a student with epilepsy. The person would not be a good teacher if those reservations were missing. Such a placement should be viewed as responsible if all possible environmental hazards have been thoroughly considered, if the student's condition has been stabilized by medication, and if there are strong indications that the student would benefit from the placement. All of us take risks each day. They are part of the rites of passage into the adult world. Placing our fears before the legitimate life experiences that a student and the special education instructor believe will be beneficial is to deny an opportunity for normalization. In this instance a disability becomes a needless handicap.

Muscular Dystrophy

When discussing muscular dystrophy, we are usually referring to one form of the disability clinically is known as the *Duchenne* type. It was named after the doctor who first described the symptoms during the nineteenth century. There are two other forms of muscular dystrophy, known as *facio-scapiluhumeral muscular dystrophy*, which affects the facial and shoulder muscles, and *limb-girdle muscular dystrophy*. These two types will not be discussed since they do not occur frequently.

Muscular dystrophy is a progressive weakening of all muscle groupings, which are eventually replaced with fatty, fibrous tissue, thus destroying the usefulness of the muscle grouping. The causes of the affliction are unknown but it is suspected to be an inherited disorder that predominately affects boys. Girls can be afflicted by muscular dystrophy but it is rare. In some cases, there is no family history related to the disorder.

The symptoms generally appear when a boy reaches three years. The boy's walk appears to be clumsy and awkward. Tiptoeing will occur as a result of weakening of the muscles that lift the foot. A classic symptom of muscular dystrophy is *Gower's sign*, which refers to a child's getting up from the floor by walking up his legs by using his hands. This symptom indicates that the thigh muscles are too weak to lift the body.

There is no known cure for this disorder. Most boys use wheelchairs by the age of ten. They eventually die in their late teens due to heart failure or a lung infection that results from the weakening of the breathing muscles.

As for treatment, there is no effective one. The goal of any treatment is to prolong the effective usage of the body's muscles. Physical therapy is used to keep a child's range of motion abilities and breathing capacity as functional as possible. Some muscles may contract and joints can stiffen. Physical therapy is used to prevent such developments. As the boy's walking skills diminish, therapy is useful in learning how to use crutches, braces, and wheelchairs. Surgery is used if any deformities of the legs or spinal cord occur.

Because the prognosis appears so hopeless, some people could become disheartened about what can be done for such an individual. Because a child's lifespan is shortened, it does not mean that a young person's experiences need be dampened. These experiences need to be enriching and filled with enjoyment, creativity, fulfillment, and satisfaction.

Other Disabling Conditions and Health Impairments

The following section consists of brief descriptions of disabling conditions and health impairments that can be present in a student population.

Hemophilia

Definition	A disorder of the blood system. Specifically, it refers to the blood's inability to clot if an injury occurs. Can affect any part of the body. Caused by a genetic trait.
Classroom considerations	The student should be able to participate in activities where injuries are less likely to occur. A shop environment where most of the work is done with machines may be an appropriate placement. Gloves would be essential for protection of the hands. Lifting heavy objects should be avoided. A leather apron should be worn in order to protect the chest.

The student needs to be given every opportunity for growth and development. All placement selections should be made with safety as a major concern, but a balance needs to be made between the student's safety and his needs for personal development.

Asthma

Definition

Difficulty in breathing caused by mucus forming in the bronchial passages of the lungs. The mucus is a result of the presence of a foreign substance in the lungs, such as dust, wood fibers, or animal hairs. It appears the affected children, especially boys, outgrow the disorder but some effects do linger, such as abnormally formed lungs.

Classroom considerations

By the time a student enters a secondary program, he may show signs of overcoming the disorder. Although this situation may exist, placement has to be considered on an individual basis. If the student is allergic to any substance that may be present, medical steps, including desensitization, need to be discussed.

Heart Disease

Definition

Normally associated with middle-aged men, some children are affected by it as well. It appears as a congenital defect (i.e., malformed vessels), or a part of the heart is malformed. Another type of heart defect results from rheumatic fever.

Classroom considerations

Strenuous activity or overexertion are the only two problems to be avoided. Where lifting becomes a necessity, the enlistment of aid from a fellow student will be the best means of overcoming the difficulty.

Spina Bifida

Definition

A congenital, neurological defect of the spinal column that generally affects an individual's ability to walk. The nerves that lead to the body have relocated themselves in a sack somewhere along the spinal column and any area below the sack is paralyzed. The sack is surgically removed shortly after birth. Bodily functions above the sack occur normally.

Classroom considerations

Because the disorder generally affects the lower half of the body, an individual's arm and hand skills should have no limitations except for the strength needed to lift heavy objects. Mobility is provided for through the use of a wheelchair.

The one possible problem area is in the student's ability to control bodily wastes. A person with spina bifida has no control over this bodily function. Boys generally use a condom-type drainage system that empties into a urine bag strapped to the student's leg. The student should be able to empty it independently. The bag presents one major difficulty—breakage—and appropriate preventive measures need to be taken.

Girls either use a catheter (a tube coming from the bladder) that empties into a bag, or a surgical procedure is employed to implant a by-pass and have the waste deposited in a bag on the side of the body.

The disposal of these waste products should not prevent a student from becoming involved in a life-centered educational program. Any difficulties should be solved before a student enters such a program. Patience and understanding will be required if a problem occurs. Otherwise, the student should not have any difficulty in becoming an active participant in the program.

Spinal Cord Injuries

Definition

A disability that results from a traumatic injury to the spinal column. The injury affects all of the body located below the point of injury. The lower the point of injury, the greater will be the individual's skills. Usually, a person's legs and lower abdomen are affected.

Classroom considerations

Although the disabilities are completely different, spinal cord injured students are similar to students with spina bina. Individuals with spinal cord injuries use wheelchairs for mobility and may require special devices for the removal of bodily wastes.

One way in which the two disabilities differ is that a person with a spinal injury may have more psychological difficulties. Until the injury to the spinal cord, the student was as healthy as any other student. All of a sudden the student has become disabled and he/she has to undergo a period of readjustment. This period may affect study habits, social interactions, and self-esteem. Being in a technical education program may provide an opportunity to rediscover him/herself, his/her potentials, and provide new goals.

Specialists and Resources

When teaching a student who has a physical disability, problems arise that require specific solutions. This chapter can only supply general information. The following section will discuss the specialists and resources that can assist a teacher and improve a student's educational experience.

Specialists

Special Education Instructor

The special education instructor received specialized training on different teaching techniques to be used with the physically disabled. He/she should be viewed as a primary resource.

- Given a specific task, the special education instructor will be able to assist in developing the best teaching method for the specific individual.
- The instructor will be able to supply information concerning what will be the best motivating technique.
- The instructor can assist in the positive introduction of the disabled student into a life-centered program.
- The instructor should be able to answer any specific question concerning a specific disability.

Occupational Therapist

The occupational therapist is trained in therapeutic techniques that enable a disabled person to gain better use of the hands and arms. A therapist has extensive knowledge in the development and use of equipment that can improve a person's arm/hand skills.

- The therapist can supply specific suggestions on how to best teach a student the use of a particular tool.
- If a student has a dexterity difficulty, the occupational therapist can devise and implement an individualized training program.
- An occupational therapist can offer valuable suggestions concerning the storage of tools and how to effectively position a student in order to ensure the most effective use of the tool.
- The occupational therapist can effectively assist in rearranging equipment.

Physical Therapist

The physical therapist is mainly concerned with a person's means of mobility—helping to gain greater control and use of the legs. As with the occupational therapist, the physical therapist will be valuable in arranging the room layout to assure greatest maneuverability.

Speech Therapist

A speech therapist has been trained in techniques of language improvement and developing alternative means of communication, such as communication boards, language cards, and signs.

- The speech therapist can develop the best means of communication for the student in a shop program.
- The therapist can also show the other members of the program how to use the student's communication system.

Vocational Rehabilitation Counselor

The primary responsibility of the vocational rehabilitation counselor is to oversee the vocational training of a disabled person. Most of the services that a counselor provides are directed toward adults, but some services can be provided for a high school student.

- If a particular device is needed for a student and if the student qualifies for assistance, the counselor through the local department of vocational rehabilitation can purchase the needed item.
- The counselor can also provide for a vocational evaluation of the student and recommend eventual vocational training facilities. Such an evaluation could offer valuable suggestions as to which prevocational skills could be targeted in a life-centered education program.

Student's Parents

By enlisting the aid of the student's parents, a teacher can obtain valuable information about the student's disability, motivation, and skills. The parents can also assist if any behavior problems arise.

Resources

The following professional organizations can supply specific information about specific disabilities and give lists of reference materials.

Cerebral palsy	United Cerebral Palsy Association of America 66 East 34th Street New York, NY 10016
	National Easter Seal Society for Crippled Children and Adults 2023 West Ogden Avenue Chicago, IL 60612
Epilepsy	Epilepsy Foundation of America 1828 L Street, N. W. Washington, DC 20036
	National Epilepsy League 116 South Michigan Avenue Chicago, IL 60623
Muscular dystrophy	Muscular Dystrophy Associations of America 810 Seventh Avenue New York, NY 10019
Spina bifida	Spina Bifida Association of America P.O. Box G-1974 Elmhurst, IL 60126
Asthma	Allergy Foundation of America 801 Second Avenue New York, NY 10017

Occupational therapy	American Occupational Therapy Foundation 6000 Executive Boulevard Rockville, MD 20852
	American Orthotic and Prosthetic Association 1440 N. Street, N. W. Washington, DC 20005
Physical therapy	American Physical Therapy Association 1156 15th Street, N. W. Washington, DC 20005
Speech therapy	American Speech and Hearing Association 9030 Old Georgetown Road Bethesda, MD 20014
Vocational rehabilitation	National Rehabilitation Association 1522 K Street, N. W. Washington, DC 20005
	Goodwill Industries of America 9200 Wisconsin Avenue Washington, DC 20014
Accessibility	Architectural and Transportation Barriers Compliance Board 330 C Street, S. W. Washington, DC 20202
	National Center for Barrier-Free Environment 8401 Connecticut Avenue Washington, DC 20015

Some of these organizations have local offices. Before writing to these organizations, check your local telephone directory.

Integration

Attitudes and General Classroom Strategies

In order to ensure the success of the special needs student in a technology education class, the steps used to integrate him or her have to be carefully evaluated. There are many points along the process of integration that could affect the student's long-range performance. All of the critical points can be anticipated and appropriate measures can be formulated. It must be remembered, however, that each placement will be different and each disabled student has different skills and needs.

The first and most critical question is: How does the student feel about such a placement? The student may have been exposed to the use of tools all his or her life, but due to a disability may only have mastered limited use; or, the student may have always wanted to learn about these tools but never had the opportunity. Thus, the student may be approaching the placement with a positive desire to learn. Even so, this positive view may be tempered by hesitancy resulting from fear about placement in a classroom with regular students. Some students may have had only limited involvement with regular students.

One method that could be employed to help the student overcome initial feelings of threat or fear is to bring the student into the classroom before the placement occurs. Such a visit would acquaint the student with the room, the other students, the tools and machines, and initial class activities. Some students may wish to talk with the class after this initial visit and, thereby, break down attitudinal barriers through face-to-face contact. An alternate method is to have an aid accompany the student on his/her first day. The aid could be an educator from the special education program or another student. Using an aid is usually more effective in helping the disabled student become more quickly integrated into the program and in developing social ties. In most situations, consulting the student or special education teacher regarding the best form of initial contact will have lasting positive effects on the subsequent placement. Some students may simply want to arrive with other students on the first day of class.

The next point of concern should be the other students in the class. Possibly they have had very little exposure to physically disabled students and would benefit by learning about the particular disability. They probably have many questions that need to be answered and misconceptions that need to be corrected. If these steps are ignored, it is quite possible that the disabled student may not become fully integrated in the class because no social bonds develop, and an opportunity to begin to incorporate a disabled individual into the greater society could be lost.

Technology Lab Strategies

Environmental Adaptations

Wheelchairs	Because of their size wheelchairs require at least five feet clearance around each piece of equipment. Such clearance may require that a shop room's floor arrangement be changed. If rearrangement is not possible, then at least one route to each major piece of equipment should be provided.
Work benches	Some workbenches are designed for standing workers. For a person in a wheelchair a table of standard height and width is needed. A semicircular cutout in the table allows the student to wheel himself/herself into the table for greater use of the tabletop.
Crutches	The major concern about the use of crutches is the condition of the floor. Large amounts of dust on the floor are hazardous, as is a newly waxed floor.

Storage	Tools need to be placed where they can be easily reached by a person in a wheelchair. Sometimes storage bins are built underneath tables that have extended tops. These tops make it impossible for a student to reach the bins, whether they are in a wheelchair or not. If such bins are present, simply cut away a section of the tabletop to permit easy access. A good place to keep tools is on the outside of cabinets, on walls, or on shelves. Two things to consider about this type of storage are, (1) keep the tools within easy reach, and (2) keep the tools separated from each other by a few extra inches. Wide placement of the tools enables a student to reach the desired tool without disturbing the others. Tools that are hung on hooks should be placed on hooks that allow the tools to be easily slid off and on.
Lapboards	If cutouts in tables are not possible, lapboards are the next alternative.
Standard tables and chairs	For those special needs students who do not use wheelchairs, working at standard tables would still be better than at raised benches. Being at raised benches would require raised chairs. Sitting on such a chair may be frightening for a student because of height. Standard chairs allow a student to place his or her legs on the floor, which increases stability. If cutouts are not possible, use chairs with armrests. Armrests provide more stability and can be used to tie straps to if a student needs added support while sitting.

Adapting Instruction

Lectures	Some students may not be able to write or to write quickly. Using someone else's notes for study can solve this.
Tests/writing	As with taking notes, writing answers to tests and worksheets may be difficult or impossible. Viable alternatives are orally administered tests and worksheets filled out with a buddy or peer tutor.
Extra material	Provide extra material for practice when a student with motor/coordination problems is first attempting to master a skill.
Demonstrations	Utilize mobile demonstration tables that can be moved near the student.

Equipment and Tool Modifications

Tools	Use special lightweight tools. Enlarge handles with padding. Add guides to tools. Guides are used for inserting material when a student has difficulty with alignment. Guides can be used for tool setup or to guide the movement of material. In figure 3-2 a wire cutter has been adapted by

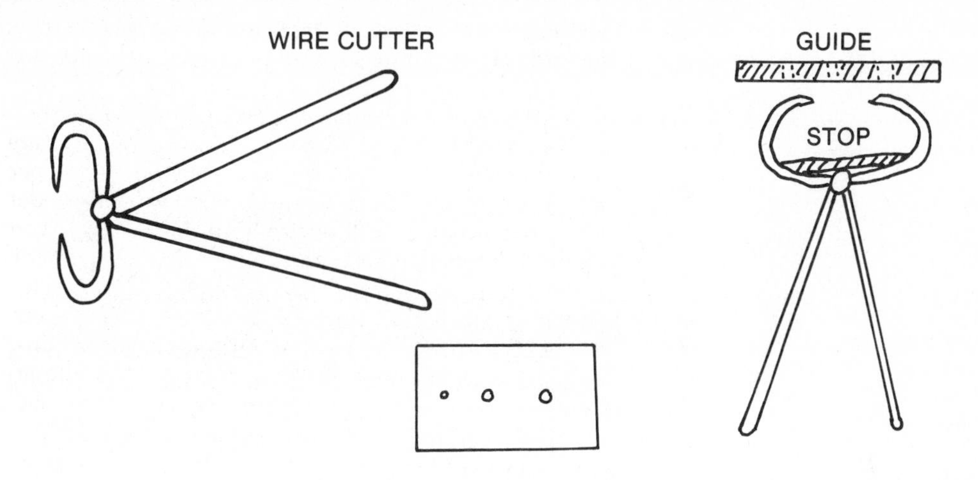

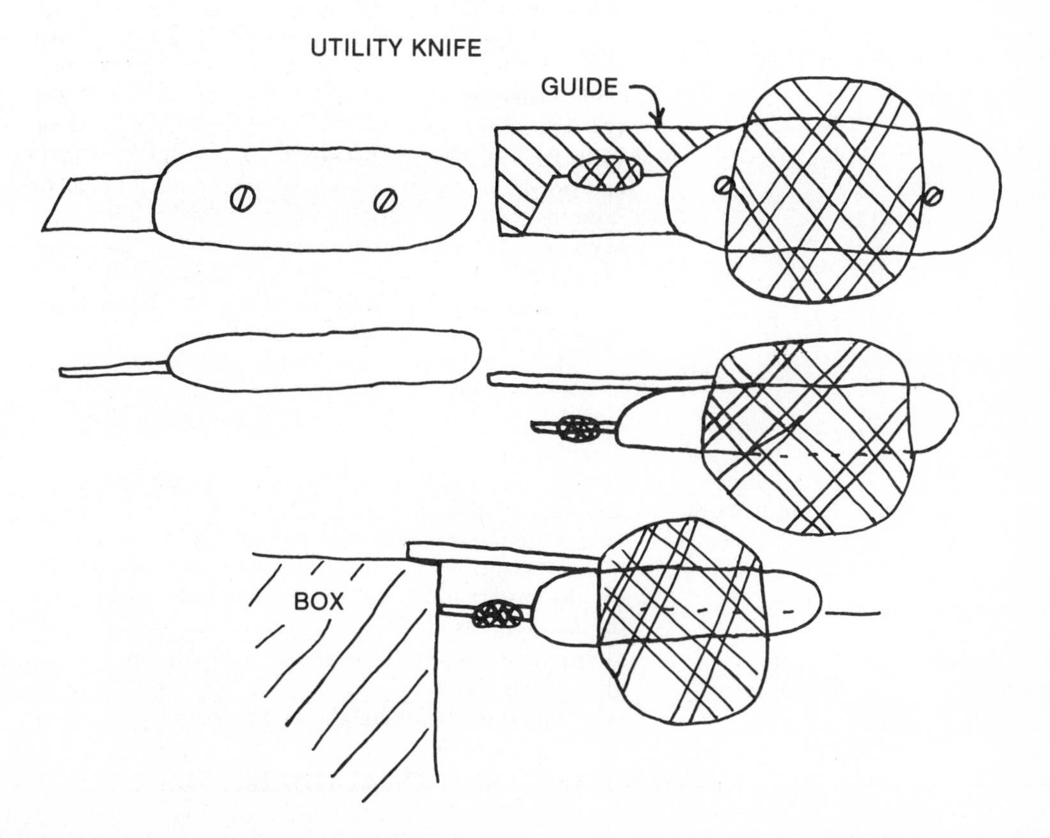

Figure 3–2. (Illustrated by Michael McHale and Todd Fly, Olympia (Washington) High School drafting students.)

placing a guide in front for inserting wire, and a stop in the throat of the cutter for ease in measuring length. A simple utility knife has been adapted for cutting open boxes. First, the tool handle was enlarged with tape for ease of grasping. Then a plastic guide was added so the blade could be run around the top of the box. A student with the use of one hand used this guide to work in grocery stores. The adapation was simple. (See chapter 4 and appendix D for additional ideas.)

Equipment — Add guards that permit visibility but add protection against sudden or unexpected spasms.

Power switches and similar controls may have to be extended in size or relocated to make it easier to control device.

Hand controls can be adapted for foot operation or foot controls adapted for hand control (e.g., foot pedal on spot welder, or sheet metal shears can be extended with a metal rod to allow for hand control).

Semistationary equipment can be located on portable tables with adjustable height.

See the Gestetner 319 Press in appendix D for many ideas on adapting equipment that can be duplicated on other machines.

Variations — Use conventional tools like work-holding devices in new ways. For example a miter box can be used to hold wood and provide additional clamping; a portable drill can be used for screwdriver operation; standard jigs and vices used in metal can be used in wood shop for holding material.

Utilize jigs and fixtures from mass production as supplemental work-holding devices.

Consult the student, who has wealth of experience at adapting objects to meet the demands of daily chores. (See Chapter 4 for ideas pertaining to rehabilitative engineering.)

Tool use consultation — The occupational therapist can provide more specific suggestions concerning specific tools. Some general adaptations are the installation of extended grips on some machines, such as drill presses; placing nonslip handles on tools with hand grips; and enlarging the grip surface on such tools as hammers.

Safety devices — The student needs to learn to use standard safety glasses, gloves, and a work apron to protect the chest. Some persons subject to seizures might wear helmets for protection.

LEARNING AND EMOTIONAL DISABILITIES

Three general impairments can produce behavior or learning difficulties: behavior disorders, learning disabilities, and mental retardation. The behaviorally disordered (or emotionally handicapped) student often has IQ scores which are average or above. This kind of student, however, usually has limited ability to benefit from full-time instruction in the regular curriculum.

The learning disabled student typically scores in the average IQ range but experiences difficulties with specific skills or subjects, usually academic subjects. As a result, this student will become angry, frustrated, or depressed at times. This can lead to misbehavior.

The mentally retarded student scores below average on standardized intelligence tests and has only limited ability to learn at the same rate as other students. This sometimes results in a display of inappropriate behavior.

These three disabling conditions are briefly described here. Suggestions for spotting potential disabilities and strategies for teaching are listed non-categorically. The non-specialist can use a single response to the learning and behavior problems exhibited by these three types of students. The teacher's goal is to develop a methodical, consistent way of handling behavior problems. It is important to consider alternative instruction sequences that might make the lessons more accessible.

Unlike the previously-described sensory and physical impairments, these handicaps usually are invisible. They are not as easily diagnosed. Remediation or adaptation of regular instruction are less simple. Students with learning and emotional impairments constitute the majority of students served in special education, however. As a result, access to experts and resources is more commonly found in the local school.

Mental Retardation

One of the oldest categories with a history of changing labels, this disability comes with a trunkload of myths and misconceptions. A simple way to view mentally retarded persons is that it takes them longer to learn or that they have a slower cognitive development. Typically, there is nothing else to distinguish people who are mentally retarded; they have the same distribution of personality and character traits as the non-retarded population. However, some myths persist, such as these:

1. Mentally retarded people look different.
   This is not true, especially in light of the fact that 89 percent of persons labelled retarded are mildly involved and are not physically different from the general population.
2. Mental retardation involves mental or emotional illness.
   This is no more true of mentally retarded people than of the general population.
3. Retardation is hereditary and sterilization will help stamp it out.
   There are two hundred known causes for retardation of which heredity contributes a small percentage.

4. Mentally retarded persons have an abnormal sex drive, are more violent, and are more apt to commit crimes.
   Sex drives among retarded people are the same as for other people. However, without training (as with anyone else), inappropriate behavior can result. As for violence, many instances of antisocial behavior among mentally retarded people often result from thoughtless persons playing pranks. It is fact that many retarded persons are often docile and have to be trained in self-assertiveness. Violent tendencies seem to have the same distribution among the retarded as among other subgroupings. The vast majority of persons with mental retardation are capable of holding jobs, raising families, and contributing to their communities.
5. It doesn't matter what you say in front of mentally retarded people because they won't understand anyway.
   Retardation does not imply hearing impairment or lack of sensitivity.

Definitions of mental retardation have changed many times and are still a source of controversy. A functional definition of "slower cognitive development" will suffice here. In a society that places great emphasis on cognitive powers (advertising, education, sports, sophisticated technology, complex social issues, etc.), it is reasonable to establish laws to advocate for the rights, independence, and dignity of citizens with mental retardation (without assuming custodial care is needed or that these citizens have nothing to contribute on their own). It is likewise reasonable to assert that without such advocacy and proper educational opportunities a person may adopt inappropriate role models, or make mistakes in judgement that are not "symptoms of retardation" but reflections of a lack of appropriate education. The very fact that IQ scores can be improved with training or that someone can be "tested out" of the "mental retardation" label gnaws away at traditional definitions and views relating mental retardation to intelligence and potential. Increasingly, we are turning to definitions of mental retardation that focus on function or adaptive behaviors. In this instance, the ability to meet "the standards of personal independence and social responsibility expected of his age and cultural group" becomes the more correct basis for definition (Brolin 1976). Definitions that only refer to IQ scores (68–52 Mild, 51–36 Moderate, 35–20 Severe, 19 and below profound) may have little meaning to a technology instructor.

Learning Disabilities

In attempting to define learning disabilities and emotional handicaps (or behavioral disorders), we find that there are at least twenty-eight different labels, among them brain damaged, dyslexic, emotionally disturbed, hyperkinetic, minimal brain dysfunction, perceptually handicapped, and socially maladjusted. All describe a common fact, that a student who does poorly behaviorally or academically is placed outside the realm of acceptability. Most professionals agree that learning problems and emotional problems have a reciprocal relationship. When school programs are in line with a student's abilities, indicators such as emotional tension, anxiety, aggressive behavior, and withdrawal tend to diminish. While medication is used, it remains controversial. Often a student classified as learning disabled scores in the normal and above normal IQ range and displays some areas of grade-level ability while two or more grade levels behind in other subjects. Some learning disabilities of this type are: dyslexia (reading disability), disgraphia (poor writing), developmental asphasia (speaking difficulty), and minimal brain damage (an umbrella term). Severe cases of brain damage, as evidenced in a trauma to the brain, are easily diagnosed since a severe loss of skills is usually evident.

Because the definitions lack clarity, the lines between mild to severe are difficult to establish. Additionally, no one knows for sure what percentage of the total population displays some learning disabilities. By the time a student enters high school, many mildly learning disabled students have discovered how to blend in (e.g.; not knowing how to tell time is merely being tardy for class). While definitions will probably remain indefinite and root causes a subject of controversy, one thing is clear: The student can profit from individualized instruction based on sound educational programs and methods.

Emotionally Handicapped

Like the learning disability syndrome, emotionally handicapped or behaviorally disordered labels cannot be neatly packaged. General criteria for this disability are very similar to learning disabilities. They include:

1. An inability to learn that cannot be explained by intellectual, sensory, or health factors.
2. An inability to maintain satisfactory interpersonal relationships with peers and teachers.
3. A generally pervasive mood of unhappiness with the tendency to develop physical symptoms, pains, or fears associated with personal or school problems.

Floy Pepper provides a functional summary by describing this disability as leading to trouble in the school, community, and family where the degree of trouble reflects the degree of severity.

Floy Pepper goes on to establish three possible degrees of severity: severely disturbed or psychotic, seriously emotionally disturbed, and socially maladjusted. The severely distrubed are rarely found in the public school and require clinical and psychiatric help in a special setting. Often they are out of touch with reality and engage in bizarre acts and other autistic behaviors. The seriously emotionally disturbed are often in special classes in the school. Basically, they cannot maintain themselves in a regular class due to poor attendance, aggressive behavior, chronic disobdience, inappropriate sexual behaviors, and the like. These students can benefit from some technology classes when their interests lie in this domain. Many of these students are quite bright and can make a real contribution. According to Pepper, the socially maladjusted "have social problems which are a reflection of the failure in the social structure." Normally, these students will not be served by special education and often come under the purview of the legal system. These students are typically involved with vandalism, illegal activities, sexual acts, runaways, truancy, and the like. (Pepper 1979). The regular teacher most likely is already instructing these students.

Potential Signs of Learning or Emotional Handicaps

It has been suggested that the definitions of these three categories may, in fact, more often reflect professional turfdom (reading specialist, mental health workers, educational specialist) rather than discrete and identifiable categories. This implies that the lay person should observe the student in terms of general behaviors and educational performance rather than attempting to look for discrete categorical symptoms. Thus, the potential referral should be to special education in general without trying to channel the child into a specific program. These are some symptoms that can only generally indicate a reason for referral and testing and do not indicate a definite impairment. In fact, many people who never experience handicaps exhibit these symptoms:

- Does not easily recall information or demonstrations.
- Exhibits immature, impulsive, or unsafe behavior.
- Is easily distracted or engages in frequent horseplay.
- Has poor language or expression.
- Is usually tense, easily frustrated, shaky, explosive.
- Imagines teacher or peer persecution.
- Demands attention.
- Is overly self-critical and has a negative self-concept.
- Does not stay with tasks, is easily distracted, and does not understand when others do.
- Hard to motivate, lethargic.
- Confuses and reverses positions, directions, symbols, numbers, spelling.
- Excessively clumsy, lacks coordination.
- Fails to grasp consequences of behavior.

Resources

Agencies that might respond with informational literature are the National Association for Mental Health,; the National Association for Retarded Citizens; the American Association on Mental Deficiency; and the Association for Children with Learning Disabilities. These agencies are listed in appendix C. Additionally, standard school specialists might be of assistance, including the special education instructor, reading/math specialist, Title I specialist, speech therapist, school psychologist, and school counselor. They can assist with possible referrals or teaching concerns.

Integration

Since emotional problems are often keyed to how students perceive their class performance, proven teaching strategies cannot be easily separated from techniques for responding to disruptive or inappropriate behaviors. Chapter 4 discusses these techniques (which can be applied to skill training, behavior management and classroom management). Chapter 5 contains specific suggestions for adapting the curriculum and instructional materials and techniques. Using these techniques to focus on a student's learning and emotional needs can effectively help him or her grow and develop.

Responding to Student Behaviors

Perhaps one of the contributing causes of teacher anxiety and burnout is the daily stress of teaching in a tense, undisciplined atmosphere. Fortunately, there is a method a teacher can use to reduce the possibility of a student's becoming emotionally disturbed or acting out tensions and anxieties. In successfully responding to one student's negative behavior, the teaching environment will be improved for all learners.

This method is based on the premise that the teacher has a methodical or consistent response to appropriate and inappropriate behaviors. The first step is to inventory your current teaching practices or styles in light of any existing behavior problems. The next step is to explore existing models for human behavior (and respond to that behavior) that will lead to the adoption of a consistent approach. No single model needs to be adopted wholesale. Some can be combined and others provide useful elements. Some of the more commonly used models in education are:

Teacher Effectiveness Training
Behavior Modification (discussed in chapter 4)
Dreikur's Theory of Misbehavior (discussed in chapter 3)
Transactional Analysis

By adapting a model to your classroom and personality you have the following advantages:

1. You have an immediate method for perceiving and analyzing behavior, which leads to a logical and consitent pattern of response. This lessens the tendency to become personally involved or feel challenged or threatened, and offers you some professional distance.
2. Your expectations for behavior can be clearly stated in advance and the consequences articulated. This reduces the possibility of feeling manipulated in a given situation and makes the student a participant in the expected classroom behaviors.
3. You learn to observe your own behavior as well as the student's. This encourages you to control your behavior before attempting to change that of the student. This is a more realistic sequence.
4. It makes you conscious of reinforcing and recognizing student efforts and problems. This conscious attention to students can eliminate many problems centered on a student's misperception that he or she needs special attention in order to feel he or she belongs.

Example

To illustrate how this can be employed, a brief overview of Rudolf Dreikurs's methods will be illustrated here. It is based on his book *Maintaining Sanity in the Classroom* (Dreikurs, Grunwald, Pepper 1971). Some of the principles underlining his view of human behavior are:

1. Everyone wants to belong. If a person feels he/she belongs, he/she maintains his/her courage and presents few problems, doing what is required and obtaining a sense of belonging by participating and being useful. Once a person is discouraged, the interest turns to a desperate attempt to achieve status. These attempts are misbehaviors, or behaviors with mistaken goals.
2. Every action of the student has a purpose of which the basic aim is to belong to a group in school or in the family. Misbehavior generally has four distinct goals that may be pursued independently or together: attention-getting, power-seeking, revenge, and assumed disability.

This basic understanding can lead to a method of responding. While Driekur's method has been simplified here, it can suggest a model for human behavior that leads to consistent teacher response. The model analyzes four common underlying causes of misbehavior and shows how the teacher can use this analysis to reduce misbehavior. The teacher's major goal should be to control his/her behavior and not the student's; a teacher can rarely win in a confrontation with someone who is pursuing a hidden logic. The response should include:

Setting the limits of acceptable behavior by stating the rules and the consequences of breaking the rules (e.g., failure to ask permission to use the power saw results in loss of lab time).

Showing a positive model for behavior (e.g. show how to belong; show how to gain status by reinforcing student's initial attempts in the class, or by designing instruction that leads to initial success).

Supporting all the student's attempts to change (e.g., verbally reinforce good behavior, offer the student the chance to decide between good behavior and misbehavior and the logical consequences).

Many people using this method have devised ways to look at misbehaviors and determine general ways of responding. The following chart illustrates one such summary:

Classroom Instructional Strategies

Teaching students with special needs does not involve a body of techniques or methods beyond the grasp of everyday successful methods. Comments from instructors in industrial arts testify to this:

"We aren't doing anything that every teacher shouldn't be doing. It's attitude" (Boland 1979).

"Don't concentrate on what they can't do, but make the situation fit what they can do" (Boland 1980).

We can summarize instructional strategies from practical experience, and from the few existing sources, in the suggestions listed below.

1. Get students involved in technology—as early in their school careers as possible.
2. Don't seat all the special learners together. Students with behavior problems often seem to benefit by being placed in the midst of others.
3. Don't treat special learners differently from the others; expect as much effort and concentration from them as you would from any student.
4. Learn the students' names and use them outside of class. The status you may lend someone this way may prove invaluable to everyone involved.

MISBEHAVIOR ANALYSIS AND SUGGESTED RESPONSES

Teacher's Feelings	Possible Goal of Misbehavior	Possible Response
	Attention-Getting	
Teacher feels annoyed or like a nag since behavior stops only when student receives teacher's attention or time.	Student shows off, horseplays, interrupts. Student acts helpless, funny, boastful, fearful.	Offer attention at the beginning of each day or lab. Ignore inappropriate behavior; follow consequences for later disruptions. Do the unexpected; use humor. Ask the student later: "Do you feel ignored? Special?" "Why do you want my attention?"
	Power-Seeking	
Teacher feels confronted, threatened. Giving attention does not eliminate the problem.	Confrontations, arguing, yelling. Rebellion, noncompliance. Inappropriate displays (e.g., breaking safety rules, inappropriate sexual displays).	Refuse to participate in power struggle. Offer an alternative behavior. Do the unexpected; use homor. Use peer pressure: "Why do you think John refuses to cooperate and is holding up the demonstration?" "Mary is upset. How can we help her?" Ask: "What is it you want?" "Do you want to be the teacher?"
	Revenge	
Teacher feels tricked, hurt, revengeful, let down.	Stealing, vandalism, bullying, refusal to clean up.	Pair the student with another student who has esteem or status. Show patience and continued trust. Discuss later: "Did you want to make me suffer the way you suffered?" "Did you want to get even?" "Did you feel wronged?"
	Assumed Disability	
Teacher feels frustrated	Student never engages in power struggles, feels hopeless, inferior, inadequate and cannot compete. Student appears withdrawn or inept; attempts to be left alone.	Don't reprimand. Always encourage. Try to find a task at which the student can excel and achieve recognition. Explain that lack of skill is not equated with lack of worth. Ask: "Do you feel you might as well be last if you can't be first?" "Do you want to be left alone?" "Do you feel 'What's the use?'"

5. Teach what is needed and useful. If math or reading or theory is too difficult, find alternatives (calculators, simplified models of theory, charts for troubleshooting).
6. Provide a longer time for mastery, if necessary; provide for repetition.
7. Simplify directions and worksheets, if necessary.
8. Introduce only one unknown at a time (e.g.: introduce one concept of metallurgy at a time coupled to hands-on practice).
9. Teach by demonstration, when possible.
10. Make instruction systematic, use task analysis (chapter 4), and supplement instruction with programmed texts, audiovisuals and pictures.
11. Teach study, reading and test-taking skills. Have students make tests in order to learn how to take them. Progress from performance-based tests to multiple choice and matching tests.
12. Continually ask questions of all the students to verify their understanding of a presentation.

The teacher may also have to develop a method for responding to problem behaviors. A general approach has already been described. These strategies will also be useful:

1. An emotionally disturbed student may not view his/her behavior as odd or wrong. Rather than discuss, sometimes it is useful to simply state rules of expectd behavior and the consequences. Avoid stating directives as questions. Rather than say, "Will you please put on the safety goggles?", state it, "Safety glasses must be worn when operating the table saw."
2. Students will often be testing the instructor, looking for loopholes in list of rules. Keep rules simple and brief. Don't overreact to infractions and feel personally threatened.
3. Students with emotional handicaps often go through mood swings with consequent improvements and reversions. Don't overreact to declines. Be supportive at all times, when possible, and encourage this in other students.
4. Students may become disruptive if they feel they cannot do the assignment. Find alternatives to expulsion as expulsion may often be the short-term goal of the student. This may involve reduced lab activities (tool room aid only) or contingency contracts tying behaviors to goals the student desires.
5. Try to establish in advance mutually agreed upon responses to crisis or noncompliance or disruptive behavior which involves special education teacher or administrators.
6. Be systematic and use routines in daily operations. Introduce change gradually; or, prepare students for change in routine by first talking about it.

7. Students may be justifiably preoccupied with personal-social problems. Provide opportunities for talking or just sitting away from others.

Technology Lab Instructional Strategies

Chapter 5 contains many specific examples of methods that can be related to adapting the curriculum and the instruction for special needs learners. These are a few general guidelines:

1. Involve as many senses as possible when learning about different materials and processes. The senses of smell and touch should not be overlooked. These types of exercises can be reinforced by incorporating competition or gaming spirit. For example, students can be asked to draw pictures of components that match electronic symbols. Students can be asked to identify different woods while blindfolded. Tape record the sounds of different machines. Ask students to listen to the tape, identify the machine, and name setup or safety procedures.
2. Provide extra time for repetition and mastery of processes. Instead of doing four different welds in four different positions, perhaps a better goal is mastering a single weld in one position. Others can be learned as time allows.
3. Strive for independence. Try to avoid demonstrating techniques using the project or materials of a special needs student. The student might begin to depend on your help. Likewise, if learning to operate equipment is part of the teaching objective, try to avoid setting up the machine. It is often better to demonstrate on scrap material and leave it near the machine for the student to use as a guide.
4. Introduce concepts one at a time. For example, introduce basic safety tools for ripping, then allow students the chance to rip before introducing crosscutting. Use projects that employ repetition (a laminated bread-board requires many ripping operations).
5. Reinforce all attempts. Confer awards for cleanup, good work habits, and other desirable practices.
6. Make sure everyone can see your demonstrations.
7. Evaluate a student only when you think he/she is at maximum potential rather than on a calendar schedule.
8. Encourage the buddy system and pair a special needs student with a nonhandicapped student.
9. Review or have the student review steps in demonstration. Never assume a lesson has been understood because the student verbally indicates understanding. Be prepared to supervise initial attempts.
10. Anticipate developmental delays in coordination or dexterity.

EXPLORATIONS

Activities

Sensory Impairments

Visit the following:
- Day school and residential school for the hearing impaired
- Companies that use teletypewriters
- State vocational rehabilitation agencies and workshops

Acquire literature about colleges serving the hearing impaired (e.g., National Technical Institute for the Deaf, Rochester, NY; Gallaudet College, Washington, DC).

Have a presentation or lecture given in sign language (role reversal).

Attend a performance given by the National Theatre of the Deaf.

Visit the following:
- State commission for the blind
- Vocational rehabilitation center for the blind
- Sheltered workshop or independent living skills center employing blind citizens

Acquire literature about specialized schools for visually impaired students (e.g., Perkins School for the Blind, Watertown, MA).

Have a presentation on mobility training given by a trainer; simulate visual impairments, wear blindfolds or dark glasses, in the technology lab.

Visit a residential school for the blind or public school program.

Observe demonstrations given on devices used to supplement sensory data:
- Optacon (tactile reading device)
- Kurzweil Reading Machine (computerized print reader with synthesized voice output)
- Versabraille (utilized for storing and accessing braille)

Construct a three-dimensional map or model for blueprints and determine how helpful or difficult it is to use.

Discuss items in appendix D and try to find examples in your community.

Physical Impairments

Visit workshops and training centers, similar to those described under sensory impairments, that train persons in daily living and vocational skills, or that employ persons in a sheltered workshop setting (e.g., United Cerebral Palsy, Goodwill, Division of Rehabilitation).

Have speakers address an audience regarding physical, social, and educational barriers (see list of organizations in appendix C).

Simulate physical impairments and examine the daily routine in your school and community (e.g., wheelchair, leg with brace, one hand tied to side).

Visit a postsecondary institution that trains persons or provides support for students in a college setting and explore their services.

Learn the fit and functions of orthoses, prostheses, and wheelchairs. (See John Venn et al, "Checklist for Evaluation, Fit and Function of Orthoses, Prostheses, and Wheelchairs in the Classroom," in references or visit a manufacturer, distributor, or hospital where devices can be found.)

Design an accessible laboratory or home in drafting. Consult *Designing for Functional Limitations* by James Moeller, Job Development Laboratory or the George Washington University Rehabilitation Research and Training Center, 2300 Eye Street, N.W., Washington, D.C. 20037.

Do a joint project with students in home economics regarding accessible clothing. A possible reference to consult is *Illinois Teacher*, 25:90, Nov/Dec 1980.

Have a discussion on terminal illness and the role teachers can play. For reference see "Supporting Chronically and Terminally Ill Children in Hospitals: A Challenge for Educators" by Uwe Stuecher (*Education Unlimited*, 2:5, Nov/Dec 1980).

Invite a speaker or driver to demonstrate ways of adapting controls on motor vehicles.

Have a discussion of rehabilitation engineering. (See chapter 4 and appendix C for references.)

Learning and Emotional Impairments

Visit workshops, activity centers, and training centers that employ persons with emotional, learning, and developmental disabilities.

Invite speakers who represent consumers or advocates in the community:

Association for Retarded Citizens United Cerebral Palsy Epilepsy Foundation

People First (consumer group for developmental disabled citizens)

Mental Health

Division of Vocational Rehabilitation

Acquire films from special education departments at colleges and universities that explore and present information on exceptional persons (e.g., Marc Gold's "Try Another Way" and "A Different Approach," distributed by the Bureau of Education of the Handicapped, Washington, D.C.

Take a class in counseling or have speakers present information from PET (Parent Effectiveness Training) and TET (Teacher Effectiveness Training).

Read and Discuss *We Are People First* by Jean P. Edwards, (Portland, OR: Ednicks Communication, Inc. 1980).

Have a panel of experts (psychologists, special educators) discuss the concepts of right brain and left brain patterns and whole brain learning and how this relates to hands-on instruction.

Readings

General

Magazines: *Education Unlimited, Teaching Exceptional Children, Mainstream*

Plays: *Take a Card, Any Card* and *The Sin-Eaters* by Martin Kimeldorf, Ednicks Press, P.O. Box 3612, Portland, OR 97208. *Stop the World I Want to Get On* c/o Sydney Stow, 5 Hillcrest Drive, Terre Haute, IN 47802

Special Class by Brian Kral c/o 2102 Perliter, N. Las Vegas, NV 89030 (All of these plays involve the realities of an educational or vocational setting).

School Shop. vol. 37, April 1978 (entire issue is on mainstreaming).

Sensory Impairments

See appendix C.

Hearing Impairments

Birch, Jack W. *Hearing Impaired Children in the Mainstream.* Reston, VA: The Council for Exceptional Children, 1975.

Doctor, P. V., ed. *Communication with the Deaf.* Washington, D.C.: American Annals of the Deaf, 1969.

Furth, Hans G. *Thinking Without Language.* New York: The Free Press, 1966.

Mindel, Eugene D., and Vernon, McCay. *They Grow in Silence: The Deaf Child and His Family.* Silver Spring, MD: National Association of the Deaf, 1971.

Moores, Donald F. *Educating the Deaf: Psychology, Principles, and Practices.* Boston: Houghton Mifflin, 1977.

Northcott, Winifred H., ed. *The Hearing Impaired Child in a Regular Classroom: Preschool, Elementary, and Secondary Years.* Washington, D.C.: A. G. Bell Assn. for the Deaf, 1973.

The reader is also directed to the following journals:

American Annals of the Deaf, published by the Convention of American Instructors of the Deaf and the Conference of Executives of American Schools for the Deaf.

Journal of Rehabilitation of the Deaf, published by the Professional Rehabilitation Workers with the Adult Deaf.

Visual Impairments

Burlingham, D. "Some Notes on the Development of the Blind." *Psychoanalytic Study of the Child,* 16: 121–145, 1961.

Cutsforth, Thomas D. *The Blind in School and Society: A Psychological Study* New York: American Foundation for the Blind, 1951.

Dillman, C., and Maloney, F. *Mainstreaming the Handicapped in Vocational Education.* Palo Alto, CA: American Institute for Research, 1977.

Lowenfeld, B. "Psychological Considerations." In *The Visually Handicapped Child in School.* Edited by B. Lowenfeld. New York: John Day Co., 1973.

Martin, Glenda J., and Hoben, Mollie, eds. *Supporting Visually Impaired Students in the Mainstream: The State of the Art*. Reston, VA: The Council for Exceptional Children, 1977.

National Society for the Prevention of Blindness. *Teaching about Vision*. New York: NSPB, 1972.

Scholl, G. *The Principal Works with the Visually Impaired*. Reston, VA: Council for Exceptional Children, 1968.

Scott, R. "The Socialization of Blind Children. In *Handbook of Socialization Theory and Research*. Edited by D. Goslin. Chicago: Rand McNally, 1969. (out of print)

Warren, David *Blindness and Early Childhood Development*: New York: American Foundation for the Blind, 1977.

The reader is also directed to the following journals:

Education of the Visually Handicapped, published by the Association for the Education of the Visually Handicapped.

Journal of Visual Impairment and Blindness, published by the American Foundation for the Blind.

Physical Impairments

Bleck, Eugene E. *Physically Handicapped Children-A Medical Atlas for Teachers*. 2d ed. New York: Grune & Stratton, 1981.

Dahl, Peter R. *Mainstreaming Handicapped Students: A Practical Guide for Vocational Educators*. Palo Alto, CA: 1979. (see volume "Orthopedically Handicapped)

Kelly, Catherine. *The Development of Individualized Supportive Services for Physically Limited Adults at a Post-Secondary Area Vocational School*. Waco, Tex.: McLemon Community College (Eric ED 146–345), 1977.

Lovitt, Sophie. *Treatment of Cerebral Palsy and Motor Delay*. London: Blackwell Scientific Publishers, 1977.

Gollay, Elinor, and Doucette, John. "How to Deal with Barriers in Schools." *School Shop*. 37:86–89, April 1978.

May, Elizabeth E.; Waggoner, Neva; and Hotte, Eleanor. *Independent Living for the Handicapped and the Elderly*. Boston: Houghton-Mifflin, 1974.

Rosenberg, Charlot. *Assistive Devices for the Handicapped*. Minneapolis: American Rehabilitation Foundation, 1967.

Silverstein, Alvin, and Silverstein, Virginia B. *Epilepsy*. New York: Lippincott, 1975.

Venn, John; Morganstern, Linda; and Dykes, Mark. "Checklists for Evaluating the Fit and Function of Orthoses, Prostheses, and Wheelchairs in the Classroom." *Teaching Exceptional Children*. 11:51–56, Winter 1979.

Walton, J. N. *Disorders of Voluntary Muscle*. 4th ed. New York: Churchill Livingstone, 1981.

Yucker, Harold; Revenson, Joyce; and Fracchia, John. *The Modification of Educational Equipment and Curriculum for Maximum Utilization by Physically Disabled Persons: Design of a School for Physically Disabled Students*. Altertons, NY: Human Resources Center, 1968.

Learning and Emotional Impairments

Birch, Jack W. and Reynolds, Maynard C., eds. *Teaching Exceptional Children in All America's Schools: A First Course for Teachers and Principals.* Reston, VA: Council for Exceptional Children, 1977.

Brolin, Donn E. *Vocational Preparation of Retarded Citizens.* Columbus, Ohio: Charles E. Merrill, 1976.

Cegelka, W. J. *Review of Work-Study Programs for the Mentally Retarded.* Arlington, Tex.: National Association for Retarded Citizens, 1976.

Dreikurs, Rudolf; Grunwald, Bernice; and Pepper, Floy. *Maintaining Sanity in the Classroom: Illustrated Teaching Techniques.* New York: Harper & Row, 1971.

Edwards, Jean. *Sara & Allen: The Right to Choose.* Portland, Ore.: Edwards Communications (P.O. Box 3612) 1976.

Edwards, Jean. *We Are People First.* Portland, Ore.: Ednicks Communication, Inc., 1980.

Furth, Hans G. *Piaget for Teachers.* Englewood Cliffs, N.J.: Prentice-Hall, 1970.

Glasser, William. *Schools Without Failure.* New York: Harper & Row, 1975.

Homme, Lloyd; *How to Use Contingency Contracting in the Classroom.* Champaign, Ill.: Research Press 1970.

Johnson, Doris, and Myklebust, Helmer R., eds. *Learning Disabilities: Educational Principles and Practices.* New York: Grune & Stratton, 1967.

Pappanikou, A. J., and Paul, James, eds. *Mainstreaming Emotionally Disturbed Children.* Syracuse, N.Y.: Syracuse University Press, 1977.

President's Committee on Mental Retardation. *Mental Retardation: The Leading Edge—Service Programs That Work.* Washington D.C.: U.S. Department of Health, Education and Welfare, MR 78.

Reinert, Henry R. *Children in Conflict: Educational Strategies for the Emotionally Distrubed and Behaviorally Disordered.* 2d ed. St. Louis: C.V. Mosby, 1980.

Scheffelin, Margaret A., and Chalfant, James E., eds. *Central Processing Dysfunctions in Children: A Review of Research.* NINDS Monograph No. 9. From Superintendent of Documents, U.S. Government Printing Office.

Shea, Thomas M. *Teaching Children and Youth with Behavior Problems.* St. Louis: C.V. Mosby, 1978.

REFERENCES

Barraga, Natalie. *Increased Visual Behavior in Low Vision Children.* New York: America Foundation for the Blind, 1964.

Birch, Jack W. *Hearing Impaired Children in the Mainstream.* Reston, VA: Council for Exceptional Children, 1975.

Birch, Jack W. and Reynolds, Maynard C., eds., *Teaching Exceptional Children in All America's Schools: A First Course for Teachers and Principals.* Reston, VA.: Council for Exceptional Children, 1977.

Bleck, Eugene and Nagel, Donald, eds. *Physically Handicapped Children: A Medical Atlas for Teachers.* 2d ed. New York: Grune & Stratton, 1981.

Boland, Sandra K. "Teacher Feature: Erline Vaughan, Martin Kimeldorf, and Steve Duncan." *Education Unlimited.* 2:28–29, 1980.

Boland, Sandra K. "Teacher feature: Four Vocational Teachers." *Education Unlimited.* 1:9–11, April 1979.

Brolin, Donn E. *Vocational Preparation of Retarded Citizens.* Columbus, Ohio: Charles E. Merrill, 1976.

Dreikurs, Rudolf; Grunwald, Bernice; and Pepper, Floy. *Maintaining Sanity in the Classroom: Illustrated Teaching Techniques.* New York: Harper & Row, 1971.

Faye, Eleanor. *The Low Vision Patient.* New York: Grune & Stratton, 1970.

Hardy, M. "Speechreading." In *Hearing and Deafness,* 4th ed. Edited by Hallowell Davis and Richard S. Silverman. New York: Holt, Rinehart, and Winston, 1978.

Harley, Randall K. and Lawrence, Allen G. *Visual Impairment in the Schools.* Springfield, Ill.: Charles C. Thomas, 1977.

Kansas Association for Retarded Citizens. *Myth and Misconception about Mental Retardation.* Merriam, Kan.: n.d.

Katz, Lee; Mathis, S.; and Merril, E. *The Deaf Child in the Public Schools.* 2d ed. Danville, Ill.: Interstate Printers and Publishers, 1978.

Maron S., and Martinez, D. "Environmental Alternatives for the Visually Handicapped." In *Implementing Learning in the Least Restrictive Environment.* Edited by J. Schifani et al. Baltimore, Md.: University Park Press, 1980.

McKay, Dixie. *Parent Handbook for Parents of Children Who Learn in Different Ways.* Salem, Ore.: Oregon Department of Education, 1975.

National Society for the Prevention of Blindness. *Teaching about Vision.* New York: NSPB, 1972.

Niemoller, A. "Hearing Aids." In *Hearing and Deafness,* 4th ed. Edited by Hallowell Davis and Richard S. Silverman. New York: Holt, Rinehart, and Winston, 1978.

Pepper, Floy. Program for the emotionally handicapped behaviors as noted on referrals for identification. From the class "Emotionally Handicapped" at Portland State Univesity. Portland, Ore.: Special Education Department, Winter 1979. (out of print).

Pieper, Elizabeth. "Labels, Language, and Self-Image." *Education Unlimited.* 2:3–4, 1980.

Sykes, Kim C. "A Comparison of the Effectiveness of Standard Print and Large Print in Facilitating the Reading Skills of Visually Impaired Students. *Education of the Visually Handicapped.* 3:97–106, Dec. 1971.

The Urban Institute. *Report of the Comprehensive Service Needs Study.* Washington, D.C.: HEW contract 100-74-0309, June 23, 1975.

Vocational Careers Program. Information disseminated in project summarizing definitions from BEH entitled: Bureau of Education for the Handicapped Definition of Severely Handicapped. Portland, Ore.: Portland State University, Fall 1979.

Williams, B., and Vernon, M. "Vocational Guidance for the Deaf." In *Hearing and Deafness.* 4th ed. Edited by Hallowell Davis and Richard S. Silverman. New York: Holt, Rinehart and Winston, 1978.

Chapter Four

SPECIAL EDUCATION TECHNIQUES APPLIED TO TECHNOLOGY INSTRUCTION

Many of the general instructional techniques used in special education can be effectively applied in a technology setting. While this chapter cannot substitute for professional preparation based on lecture, observation, and guided practice, it can indicate specific ways in which special education and technology instruction can be blended. Special educators can begin to see here how their techniques can be translated into another field, another setting. Technology teachers can gain greater confidence through knowledge of special education concepts and feel less intimidated by the aura of "specialization" expressed in such terms as task analysis, behavior management, environmental cues, and prescriptive teaching.

Many techniques associated with special education will be found in most regular teachers' repertoire. Special education consists of the systematic analysis and application of these techniques. Becoming systematic and consistent in instructional delivery and design is the process of specializing one's instruction. Individualizing the specific instructional format based on the student's uniques way of learning constitutes specializing the educational process.

The techniques described in this chapter can serve both mildly handicapped and severely handicapped learners. It should not be implied that because a learner requires specialized teaching methods that he or she cannot benefit from a regular class setting. It is assumed that when a student warrants these specialized instructional methods that extra help will be available both to the student and the regular teacher. Most likely, the students who will most often warrant this form of specialized help will be those with moderate to severe disabilities.

Today's specialized methods are based on behavioral science. Inevitably, behavioralism will undergo a metamorphosis. However, the technique contained in its core will continue to be useful and can be adapted to a variety of instructional methods. This chapter first discusses general behavioral principles and methods, then specific techniques for instruction.

These topics are followed by more specialized techniques and descriptions of programs related to serving students with severe handicaps. The criticisms of behaviorism are also noted and an alternate method is presented.

GENERAL PRINCIPLES OF BEHAVIORAL INSTRUCTION

There are many theories on how we learn. The behavioral approach is often at the center of many learning theories in special education. This philosophy has been translated into techniques called task analysis, behavior management/modification, and prescriptive and precision teaching. It is important to understand how these techniques can be used in an entry-level program or be incorporated into the individual education plan of a student in a regular class setting.

Behaviorism looks upon learning as adapting to one's environment. It is not concerned with what lies beneath the conscious observation of phenomena. In other words, it looks merely at how an intelligent organism reacts or responds to the stimulus in the environment rather than what are the unconscious or subconscious motives and causes of the act. In this respect, behavioral techniques are most easily mastered by the lay person than techniques that require training in counseling and psychology.

Behaviorists believe that we learn through association. Some of this association is accidental or incidental. Misbehavior is often a product of incorrectly associating an inappropriate behavior for a desired reinforcing event. For example, a student continues to engage in horseplay because it results in the desired reinforcer: the teacher's attention. If the instructor can provide or create an alternative path for this reinforcer, then the student may modify his behavior. This clinically stated theory can be illustrated with an example from the classroom.

Suppose a teacher encounters a student who continually demands attention. Every time the teacher gives a demonstration the student acts up to draw attention to himself. The teacher continues to berate the student, but the situation only worsens. While one could theorize that the student feels insecure, neglected, or threatened, the behaviorist would not delve into root causes. Rather, the behaviorist would only state that the student found the attention he received stimulating, motivating, and thus reinforcing. This might seem odd at first, that a teacher's reprimand is reinforcing. However, one can imagine the situation of a student who feels defeated and unable to receive recognition. The student might further associate attention in general with recognition and status. Thus, unable to distinguish himself on tests or in class performance, he seeks attention along another route, the route of misbehavior. In this instance, the teacher's reprimands have become incorrectly (in the mind of the behaviorist) associated with status. The behaviorist seeks to break down this association and establish a new one with more cooperative behaviors. This will involve changing the behavior of the teacher first, and the student second.

First, the behaviorist will ask the teacher to try to ignore the student's intrusions on the demonstration, if they are minor. If this fails to extinguish the disruptions, then the teacher will have to spell out the expected behaviors along with contingent rewards and punishments. At the same time, the teacher must pursue a course of action that reinforces the student's correct behaviors. This might consist of paying more attention to the student when he first comes in and/or finding little things in which to praise the student (such as appearance, safety, concentration, actual work). In this case, the teacher attempts to create a new association with a new behavior while breaking the association with the old undesired behavior. We have attempted to organize the environmental cues and stimuli, to reinforce certain behaviors and thus remove some of the previous randomness or incidental quality of previous associations/adaptions. Thus, a new behavior is learned.

Other principles that have evolved from the application of behavioral science to educational methods concern the importance of positive feedback and of direct associations built around repetition. An important behaviorist principle is that learning only takes place when a person is receiving 80 percent positive feedback associated with his actions. If this outlook is adopted, it will have impact on the design of your curriculum. For example, it may necessitate breaking down your introductory instruction into easily mastered units. Ease of mastery eventuates praise and leads to positive self-concept. This is finally expressed as increased motivation, which leads to further learning. The corollary to this is that punishment or negative experiences retard learning and require about four positive events to balance the impact of each negative experience. For example, when examining a student's welding progress, your criticism of the bead should be tempered with praise. Instead of saying, "too much splatter—you went too fast and you need more amperage," you increase the student's motivation with comments like, "I notice you've really stuck to it, worked hard, and have practiced the safety rules. Your bead is pretty straight. You can get rid of these ripples if you slow down a little and turn up the amperage. Welding is fun, isn't it? Keep up the effort."

Summary of Behavioral Methods

Behavioral approaches can be used to teach specific psychomotor skills as well as personal behaviors. Generally, the attempt to teach a specific skill is termed *behavior modification,* while the attempt to change behavior is often called *behavior management* (although the two terms are often used interchangeably). A brief summary of behavior-based instruction is listed on page 106 in a sequenced format. It relates to instruction for both personal behaviors and psychomotor skills.

GENERAL SEQUENCE OF BEHAVIORAL INSTRUCTION

Step 1: **Observe and Pinpoint Skill/Personal Behavior**

Skill: Task analyze the skill (break it down into parts).
Behavior: Determine the events that occur before and during the incident.

Step 2: **Baseline**

Determine the proficiency of the student or the frequency of the behavior.

Skill: Pretest to determine student abilities.
Behavior: Observe and record frequency of undesired behavior.

Step 3: **Determine Your Goal and State it in Behavioral Terms**

Attempt to involve the student in establishing a goal (Discuss concerns and expectations to arrive at a mutually agreed upon goal when possible).

Examples:

Skill: Student will strip a No. 12 wire using standard wire strippers with ends 1″ long and no errors.
Behavior: Student will begin working in class within five minutes after the students enter the lab. The teacher will supply only one prompt.

Step 4: **Design a Program**

Attempt to involve the student. When possible, mutually arrive at rewards and punishments (or changes) contingent upon performance.
Some instructors use contractual agreements.

Examples:
Skill: Determine method, content, cues, and techniques. An example of a prescriptive format is found on page 000.
Behavior: If student is able to start on his/her own and achieve the goal, then he/she, will be instructed in the use of the blueprint machine. If the data indicates nothing has changed, a conference will be held between the student and the instructors involved. If the student increases negative behavior as indicated by data, then a change in placement will be recommended at a subsequent parent conference.

Step 5: **Run the Program and Collect Data on Student Performance**

In many instances, student can collect their own data as this may be rewarding. It also shifts the burden of responsibility onto the student.

Step 6: **Periodically Evaluate Data and Make Changes**

The periods are determined according to rates established in the baseline (step 2).
Many different techniques can be used to display data for purposes of evaluation (graphs, charts, etc.).
If success is found in achieving goals, then the program is faded out gradually assuming that the student maintains mastery of the skill or new behavior.
If there is no change or a decline in positive performance, then a return to steps 1 through 4 is recommended.

Teacher Qualities Related to Behavioral Techniques

Teachers wishing to use behavior modification must be prepared to practice systematic observation and consistency in instruction. Observations are used to collect data on the success of a program or to document the need for change. Intervention is planned in accordance with the data collected. The personal qualities needed include at least average organizational skills and a willingness to adapt teaching methods and style. Even if a teacher is uncomfortable with behavioral techniques, the training will sharpen his/her powers of observation. Overall, the behavioral approach is similar to programmed forms of instruction.

The person considering behavioral approaches should be aware of the criticisms leveled against them. (see page 125). Otherwise, he or she may fall into the trap of taking a single course and applying the methods mechanically without regard to the particulars of a specific situation. Likewise, classroom constraints for a regular teacher must be recognized. The ability of a regular teacher to apply focused behavioral techniques and collect data is compromised unless some additional help is provided. However, once the techniques become commonplace, the regular teacher can find ways to support a behavioral program for a special needs student.

BEGINNING TECHNIQUE— TASK ANALYSIS

Task analysis is the first step in planning a behavior modification program. It is similar to lesson planning for regular teaching. Industrial arts teachers use a form of task analysis when they create job, operation, or instruction sheets.

Task analysis is simply a breakdown of the skill to be learned. We start by listing our objectives in behavioral terms. This means we need to state a condition, a behavior (or action), and a criterion for completion. Here is an example.

Condition: Given the necessary materials and equipment
- a. type, chase, plastic matrix, paper, soapstone
- b. molding, rubber cement, oven

Behavior: the student will prepare a rubber stamp
- a. compose type in a chase
- b. create a rubber impression
- c. mount it to a molding

Criterion: that accurately prints the text with 100 percent reproduction.

Task analysis involves a breakdown of the skill and attempts to inventory the needed materials and skills. There is no one correct procedure except one that allows the student to learn. Task analysis can be used to supplement regular instructional modes when the student is unable to benefit from regular planned instruction. For example, the regular method of imparting information might involve the following job sheet related to a rubber stamp demonstration.

RUBBER STAMP JOB SHEET

Follow these instructions for making a rubber stamp. Remember safety instruction when using the oven and hand tools. If you have any questions, ask the teacher or aid for help.

1. Compose type and lock it in a chase. Make sure it is firmly locked.
2. Place plastic matrix over chase and cover with paper.
3. Slide into press (oven) for preheating for one minute.
4. Raise bed of press to begin heating for ten minutes.
5. Remove and ply matrix from type form.
6. Dust matrix with soapstone.
7. Cut stamp gum allowing for a margin of ⅛" larger than matrix.
8. Put gum with paper over matrix.
9. Put in press for six minutes. Raise press bed.
10. Remove and strip vulcanized (cured) rubber from matrix.
11. Trim rubber and cut molding to size.
12. Attach molding to rubber with rubber cement.
13. Sand ends of molding and stain.
14. Make a print and use this as a label on your molding. Be sure to mount it so that you print right side up.

After the demonstration the student has trouble with the process. Perhaps he or she does not read or recall the correct procedure. Upon closer examination of the job sheet we find that several steps introduce new words, include measuring, refer to previous information, or combine several single operations in one step, as in step 5. It is important also to recognize that technology-based education for poor readers presents difficulties with such technical words as *matrix, preheat, ply, vulcanized, cured, compose,* and *chase*. The student may know what you did in the demonstration but require some form of supplemental instruction for mastery and usage of this new vocabulary. Often, words are used interchangeably. For example, in step 2 the matrix is placed over the *chase*, while in step 5 one must ply the matrix from the *type form*, which is a more specific description of the chase parts. In step 8 one places the *gum* over the matrix, but when it is removed it is called *vulcanized rubber*. Step 1 assumes the student recalls how to use a chase from a previous lesson. This apparently simple job sheet is not so simple after all. From vocabulary to individual steps, there are many pitfalls for the student who lacks previous experience or confidence or who experiences learning difficulties.

This job sheet can become the basis of a more detailed and simplified breakdown where the tasks involved are further analyzed. The instructor may decide that the behavioral objective can be broken down into three separate learning objectives corresponding to the three performance indicators listed under the behavior part of the overall objective. There is no end to the detail to which the task analysis could be broken down. The goal is to find the appropriate level of breakdown for that particular student or potential group of students with a similar disability.

This same job sheet could be broken down into forty to fifty steps, each step accompanied by the materials and skill required. Some students can learn directly from this task analysis sheet by themselves or with the help of an aid or supplemental instruction. Other students with a more involved disability will benefit from this task analysis when put in a more specialized program. A sample of this original job sheet modified by a task analysis is presented on pages 110 and 111.

The task analysis provides both the teacher and learning specialist with some additional information. First, it suggests four distinct parts in the process that could comprise instructional units. The length of time may be longer if further breakdown is required. Secondly, by listing prerequisite skills, supplemental and remedial instruction can be offered in the resource room to aid the student in completing the project. In this way, time is not taken away from lab time. Supplemental instruction may involve a review of procedures, names of tools or parts, or practice skills. Remedial instruction may include the use of a timer, measuring implements, and vocabulary.

Upon further examination, other information can be gleaned that may assist in the task analysis. Notice how step 1 in the original Rubber Stamp Job Sheet has now been expanded to 11 steps! This ensures review of a previous lesson and success in the initial steps. Once procedures are repeated, they can be grouped so that the student gets used to handling multiple directions in a single step. For example, steps 22 and 23 are repeated and combined in step 34. Measuring has also been eliminated for most practical operations. Steps 25 to 28 comprise the act of allowing for a ⅛" border. Instead of requiring the student to learn to measure to ⅛", which could subvert or delay the lesson unnecessarily, spacers that the student can trace around are used. Industry would use a similar technique in the interest of speed.

When first working with task analysis, the instructor need not feel compelled to jump from a fourteen-step job sheet to a forty-six-step task analysis. Start by looking at the job sheet and think in terms of skill components, or merely observe what the student has trouble with. The job sheet might be slightly expanded to represent a twenty-step task analysis. Another approach is to develop a pretest using five to ten sample skills to determine which areas should be expanded. Eventually, it must all be summarized if it is to take the form of a planned intervention that includes evaluation of student progress.

TASK ANALYSIS SHEET

TITLE: Rubber Stamp

BEHAVIORAL OBJECTIVES: Given the appropriate materials, the student will prepare a rubber stamp that prints an accurate text with 100 percent reproduction.

SAFETY PREREQUISITES: Use of oven and first aid for burns. Wash hands afterwards.

Steps	Prerequisite Skill	Materials
Part I		
1. Identify text you want to print.	Reading	
2. Obtain type and composing stick.	Use of type, fine motor skills	Type case and stick
3. Compose type.	Reading backwards	
4. Slide type from stick onto composing galley.		
5. Tie up type.	Making knots, two-handed dexterity	String
6. Untie and proof the composition.	Use of tweezers to correct/change type, reading	Tweezers
7. Correct it.		
8. Secure furniture and form.	Measuring width and length	Metal furniture quoins (for lock), frame, ruler
9. Place quoins and furniture in place.		Quoin key
10. Tighten and keep level with planer.		Mallet, planer
11. Test for lift.		
Part II		
12. Preheat oven for two minutes at 300° F.	Reading oven dial, time usage	Oven, timer
13. Trace outline of matrix on paper.	Tracing	Paper
14. Cut out paper.	Use of scissors	Scissors
15. Place matrix over type, red side down.	Color discrimin-ation	Matrix
16. Place paper over matrix.		
17. Make sure oven bed is down, fully rotate knob counterclockwise.	Directions; clock-wise, counterclock	
18. Holding chase by handle, insert by laying it on the ovenbed.		
19. Preheat for one minute.		
20. Raise bed until it tightens, stops turning clockwise.		
21. Time for ten minutes.		
22. Lower bed and remove.		

Steps	Prerequisite Skill	Materials
23. Place on unpainted surface and allow it to cool for 15 minutes.		
24. Using fingernail, slowly pry matrix from the form.		
Part III		
25. Surround matrix with ⅛" spacer.		Spacer
26. Trace this over carbon paper.	Use of carbon paper	Carbon paper
27. Transfer this shape to rubber by cutting out and tracing.		
28. Cut out rubber.		
29. Dust matrix and rubber with soapstone, shake off excess.		Uncured rubber, soapstone
30. Place matrix on vulcanizing tray, powder side up.		
31. Place rubber over matrix, powder side down.		
32. Place paper over rubber.		
33. Place in oven, raise bed for six minutes.		
34. Lower bed and remove, place on unpainted surface for fifteen minutes.		
35. Test rubber by sticking nail in rubber.		
36. Did fingernail mark disappear? If not, get instructor.		
Part IV		
37. Trim rubber under teacher's directions.		
38. Hold rubber up to molding and mark off length.		Molding
39. Secure saw and cut molding on mark.	Safe and accurate use of saw	Saw
40. Put glue on molding side that matches the rubber size.		Glue
41. Mount rubber, flat side down.		
42. Let dry.		
43. Using fine sandpaper, sand ends of mold.	Sanding	Sandpaper
44. Stain ends of molding.	Staining	Stain
45. Test out on ink pad and paper, cut out one of the impressions.		Ink pad and paper
46. Mount with tape on handle so it indicates correct way to hold stamp.		Tape

Pretesting and post-testing can assist in many ways. Such testing pinpoints the need for special assistance or instruction. Sometimes, just working for a longer period with some assistance is all that is required. Construct the pretest on a simple-to-hard continuum and do not neglect the related academic skills of reading, measuring, and technical vocabulary as these can be taught by others elsewhere. Let the student try the test items more than once—three times, in fact. Make sure the test is not based exclusively on reading or vocabulary skills. One approach is to use pictures or a demonstration of the test items. The pretest can later become the basis for an instructional program in the resource room. When taking data on the student's performance, try not to give information as to the correctness of the response. Otherwise, the student will give up or merely work for an encouraging response. If the student refuses to do a certain step, either come back to it later or return to any step of which you are unsure.

Pretesting logically leads to post-testing. Together these tests offer a viable alternative to traditional grading. This will be discussed in more detail in chapter 5. At this point, it is important to keep pre/post-testing in mind as you implement task analysis. For example, your criteria for the behavioral objective could be based on improvement rather than specific skill mastery. The objective might read as follows:

> Given the necessary materials and instruction on the construction of a rubber stamp, the student will be able to do 50 percent more of the steps on an independent basis as indicated by a pretest and a post-test.

Post-testing will also serve as a review lesson if administered periodically.

SPECIALIZED INSTRUCTIONAL TECHNIQUES

The focus now shifts from initial planning and assessment to steps to determine which specific instructional techniques the learner might profit from most. When these techniques are coupled with a task analysis, a completely individualized plan results. The virtue of a consistent and methodical format for instruction is the ability to reduce distractions in the learning process. If vocabulary, mood, style, and sequence vary from day to day, the learner may soon stray from the task and lose the concentration necessary for mastery of the content.

Overteaching

The first general area of technique could be called overteaching. This means the information you present has a certain redundant quality. This slows the pace and allows numerous experiences before mastery or some degree of competence is expected. The principle follows the premise that after many trials (a quantitative involvement), the student will make a leap into mastery (a qualitative progression). This is in contrast to requiring prerequisite skills or conceptual understanding prior to application. It is basically the principle of hands-on education, learning by doing, extended to incorporate a remediative quality. The most common example that many people have experienced is learning to type. One starts with a quantitative involvement: being drilled on individual letters. The letters in isolation soon become drills in spelling out words. After a period of time, one stops "thinking" in terms of individual letters and progresses to words as speed is increased. This is similar to any psychomotor skill. For the technology teacher, it might be most apparent in learning to weld.

Chaining

The most common technique in overteaching is *forward chaining*. Simply stated, this technique proceeds from a task analysis and each step is mastered sequentially. Certain difficult skills may be taught in isolation. Usually, the sequence or chain is broken down into instructional units. The rubber stamp has four such units. The major drawback is having to wait to experience the whole task while mastering one step at a time. Motivation often suffers as the student loses sight of the goal. For example, requiring measuring ability before proceeding on a wood or metal project may kill the interest of many students. Learning first about the many types of engines, combustion systems, and ignition systems when the student is desperately wanting to do a simple tune-up on the family car or lawn mover can be a mistake.

Reverse chaining is a modification with advantages when teaching skills have a long list of steps. For example, the task analysis for the setup procedure for an electronic meter remains the same as in forward chaining. The difference is that the student starts with mastery of the last step first. Thus, the student is assisted through the entire setup procedure. Upon reaching the last step, the student is required to do it independently. Then he or she works backwards, one step at a time, toward mastery of the entire process. In this case the student experiences the entire setup procedure each time. The student can see where he or she is going and, by the time the student is ready to do it solo, he or she has experienced the entire procedure many times over. A sample of this type of program is included on page 114. First a task analysis is presented. This is followed by a data collection sheet on the student's progress. The crucial difference is that the steps are stated in reverse order for data collection; this does not mean one teaches in reverse. In taking data, a 1 means mastery, 0 the opposite, and Ø means student needed help. A student receives no more than three trials per step. When he or she can do the step correctly three consecutive times (over three days or total trials), he is considered to have met the criterion (MC) for that individual step.

TASK ANALYSIS

TITLE: Setting Up Ohmmeter

1. Secure meter with red and black leads.
2. Turn selector to *ohms*.
3. Plug black lead into *ground*.
4. Plug red lead into *ohms*.
5. Turn range to appropriate scale (i.e., Rxlk).
6. Clip two leads together.
7. Adjust *zero* thumb wheel so needle reads 0.
8. Unclip probes.
9. Adjust *ohms adjust* thumb wheel so needle reads 1000.
10. Repeat steps 6 to 9 once.

DATA-LESSON PLAN

Step	Event Mastered	Date	M	T	W	TH	F
		Initials					
1.	Adjust *ohms adjust* thumb wheel for 1000.						
2.	Unclip probes.						
3.	Adjust *zero* thumb wheel for 0.						
4.	Clip leads together.						
5.	Turn range to appropriate scale.						
6.	Plug red probe into *ohms*.						
7.	Plug black probe into *ground*.						
8.	Turn selector to *ohms*.						
9.	Secures meter and red and black probes.						
10.	Repeat steps 1 to 4 once before finishing (this step put at end since mastery of other steps would be taught first).						

Successive Approximation or Shaping

Successive approximation or *shaping* can be used effectively in exploratory classes. In this instance, the student is not expected to perform the task independently in its entirety. Rather, the student is asked to attempt an approximation of the entire task. As the student masters the approximated task, he or she proceeds to greater competence. Essentially, it is a way of eliminating prerequisites that would become barriers to participation in the lesson. The tremendous advantage is that it allows the student to experience the entire task. The student takes on an "apprentice role" or is assisted without being demeaned. For those instructors wishing to assert more control over the initial experiences of their students or who are concerned about safety, this is an apt technique. An example might be the learning of machine tool operations in a metals class. An annual goal for the student in the individual education-plan might be the following:

> Given assistance in machine setup, safety, and operation, the student will produce required projects at his/her own pace being evaluated on attendance, work habits, attitude, and completion of the project.

In this example, suppose the student was making a small hammer and that the handle required knurling. The teacher, student aid, or another student might assist in setting up the knurling tool. The handicapped learner might center the handle in the lathe and have it checked by the aid. Then the aid would initially engage the knurling tool and let the learner run the carriage down the lathe the length of the handle. In this manner the student being assisted has been involved in the overall task and has been in charge of certain facets.

Another example of shaping might occur when teaching someone to strip wire. One might start with a tool that allows the student to merely insert the wire in an automatic wire stripper. Then a tool with a preset wire-size opening could be used as the student learns to develop a "feel" for wire stripping. Finally, different size wires and a generalized wire stripper could be used.

Assistance

Another general technique could be labeled assistance. The first level of assistance involves *prompts*. This form of assistance is used to help someone who already understands the task. The prompt helps to initiate action or a decision that will lead to the desired action. Prompts are often given students when they are tuning an auto or a radio receiver. A student may be using a test instrument but need a verbal prompt for the first time to recognize the optimum tuning point. A more common example would occur when a student is being taught to solder. In this case the student places the soldering iron on the wire joint and the instructor prompts him/her to add the solder, then prompts the student to remove the hot iron and solder. Prompts in this case could be verbal as in "add the solder *now*" or just "solder." The prompt could take the form of pointing a finger or a touch on the shoulder, wrist, or soldering iron. Since prompts provide additional information that does not normally exist in the soldering process, they are sometimes called "redundant cues." Sometimes these cues are used in industry, as in the case of color coding machinery. Here the colors indicate hazardous locations on the machine and are redundant when considered in the light of the machine's functions. You might extend the use of color to indicate sequence on machinery. Certain hot colors could signify cutting edges, neutral colors clamping, and cool colors guards as the student follows a continuum of warm to cool in machine operation.

Prompts may also take physical forms, for example, a stop on a lathe or other machine. Here, a visually impaired student could use the stop for placement of material before cutting, for control of travel of the workpiece, or cutting tool, or both. Another prompt might be raised or embossed markings on a machine to aid a student who is visually impaired. These prompts can help the student determine when he/she is approaching the cutting edge of a saw blade. A hearing-impaired student might need a visual prompt for an operation that normally depends on sound perception. Examples include an oscilloscope to illustrate engine tuning or rpm meter, and an oscilloscope to illustrate frequency filtering in electronics. Some shops use a flashing light to assist the student when the fire, emergency, or cleanup bell is rung.

Assistance can also take the form of your *physically assisting* someone through a task, similar to prompting. Some forms of this technique are called *hand-over-hand*. Often this technique is employed when a sense of the right "touch" is needed and other cues are missing. Thus, you might physically assist a student to start a hand-sawing operation. Here, you assist in developing even, continuous, and straight strokes. You may physically assist a student to "strike an arc" in welding or to "puddle" with a gas welding technique. Techniques other than hand-over-hand include adapting handles so that you do not put your hands directly on or under a student's. For example, you could add an additional handle to a saw or sheers to "piggyback" your assistance. This allows the student to grasp the tool directly and experience the movement. It is also easier to fade out your help as you gradually release the handle.

Discrimination

Discrimination represents the last area of technique. These skills are used all the time in technical settings. In some settings hearing impaired people use these skills to play music and visually impaired students learn to use color. There are two basic types of discrimination. The first is called *discrete* discrimination. Here, one discriminates between entities that have their own identities. A good example of this would be the ability to sort and utilize different electronic components. The other type of discrimination involves a *continuum*. Here, one discriminates between differences in kind or degree. This might involve finishing the surface of metal, wood, or plastic to different degrees of smoothness or polish.

When teaching someone to discriminate between discrete items, one particular method has proven very efficient. Rather than present the student with all the different components and ask that each be memorized, a different approach is used. It involves teaching each component in isolation and then adding a "distractor" or second item to see if the student can discriminate between the desired object and the distractor. Here is a sample program you might use in teaching two different components, a capacitor and a resistor.

1. Present student with a single resistor. Student repeats name until he or she can recall it independently.
2. Introduce a distractor that is in sight but off at a distance. It might be a transformer. Student again repeats the name of the item.
3. Introduce different size or color resistors with distractor moved closer.
4. Introduce random resistors *with* the distractor while the student repeats and chooses the correct item.
5. Introduce a capacitor in the same manner as steps 1 to 4.
6. Introduce a capacitor and a resistor at the same time with student identifying them correctly.
7. Introduce different sizes or colors of capacitors and resistors.
8. Repeat steps 6 and 7 with a distractor.

Obviously this system will be reduced if the student has quicker initial responses and recall.

Teaching a student to discriminate between items on a continuum is somewhat similar. Suppose the instructor wanted a student to learn different wood finishes or wood types. Start the instruction with two extremes, such as an unfinished rough-cut piece of wood and a piece with a high-gloss finish. Or, one might use a light-colored, lightweight piece of wood like pine and a piece of dark, dense hardwood like ebony or walnut. Each item may have to be introduced separately, as in the program for discrete discrimination, then the middle values are gradually introduced (e.g., a sanded and sealed piece of wood, or a piece of mahogany). Still later, finishes or wood samples requiring finer discriminations could be added. This same technique could be used to teach measuring and is similar to successive approximations.

Troubleshooting

If the student still is unable to progress, the teacher can begin by going back to reflect or observe the educational program. In order to troubleshoot the instructional program one must attempt to isolate the one or two variables that may have caused difficulty. The following list can serve as a guide to analyzing an educational program that has not succeeded in attaining a reasonable goal.

1. Is the step too large for student? This sort of flaw is the most common type.

   Branching and breaking down the step is a minor adaptation.

2. Has the program been delivered in a consistent manner?
   Perhaps new words, new cues, and new names have been used, or demonstrations and assistance have varied. This might be indicated by the fact that the student cannot seem to remember from lesson to lesson but does well within each lesson. In this case, variety in instruction has sidetracked the student. Perhaps a very detailed program that can be repeated each time is warranted. This type of program is discussed in the next section.
3. Is the manipulation in a particular step too difficult or demanding in terms of coordination or precision?
   A guide or holding device may be warranted.
4. Does the task require a nonverbal cue that the student is not able to discern?
   An example of this is stripping wire or tightening bolts on a wheel where no visible sign indicates the proper degree of tightness. Perhaps physical assistance would help. Sometimes a *mass trial* that lets the student repeatedly experience a specific skill will help. After several trials he/she may be able to do it alone.
5. Does the task require good recall or memory abilities or ability to sequence properly?
   Parts can be laid out sequentially so the student merely moves down a table to assemble something. Later this can be faded out. Another approach is to use mass trials again.

Fading Out Specialized Techniques

It is important to conclude with a statement regarding independence. The teacher must always be conscious of the need to fade out special techniques when possible. Otherwise, the student is hampered by becoming dependent upon these techniques. Several common ways of fading out specialized cues are listed here.

1. Withdrawal of the cue entirely, especially if the original adaptation was minor.
2. Planned reduction in the intensity of the cue. This can be done in the following format. Suppose a student was working on a project involving assembly from a pictorial (or blueprint in other shops). You could use the following reduction for a mass-produced item:
   a. Student assembles object with pictorial assistance.
   b. Pictorial is reduced to simple line drawing.
   c. Line drawing becomes dotted or dashed lines.
   d. Drawings eliminated entirely.
3. Reduce the number of times you give cues or assistance.

4. Introduce a delay in offering cues or assistance so that student begins to initiate the desired behavior.
5. Speed up the actions of the student so he/she will not have time to wait for cues or think about what he/she doing. This is done after the skill has been fully mastered.

BEHAVIOR MANAGEMENT

Behavioral methods can also be applied to the modification and improvement of personal or social behaviors that interfere with a student's ability to successfully interact or participate in a regular class. Behavior management attempts to teach a student a new behavior, reduce or extinguish any undesirable behavior, and increase or maintain a desirable behavior.

Students may be placed in a regular class or programs may be set up in entry-level classes that require a behavior management component in the IEP if the student is to be maintained in the class. For this reason, the technology instructor should possess a general awareness of this body of technique. This section will attempt to illustrate various facets of behavior management by using typical examples that the instructor may encounter. The maintenance of a behavior management program is not difficult. Peer tutors, paraprofessionals, and students participating in a behavior management program have all demonstrated the ability to maintain or run a program. The instructor who wishes to pursue these techniques in a regular class is encouraged to take professional courses and consult with education specialists when attempting to design a new program for a student.

The examples cited here are stated in general terms as case studies. This leaves open the option of applying this technique generally across the board for classroom management, or individualizing it to particular behaviors of specific individuals. In either case, the goal is a systematic approach toward improvement of student behaviors. This general method makes the following procedures mandatory for the implementation of any behavior management program:

1. Clearly state the expected behaviors or accepted rules for participating in class. These can be related to safety, work habits, discussion behaviors, or personal behaviors.
2. Establish a system of rewards related to fulfillment of these expectations. These can include field trips, time in the lab, work experience positions, and independent projects.
3. Establish in advance a consequence or levels of response to inappropriate or undesirable behaviors. Examples of this are covered under noncompliance.

When possible, students should be involved in the setting up of a classroom behavior management system to ensure their participation and understanding as well as making the process democratic and, thus, fair.

The instructor should start by asking the question, "What measures can I take in advance to reduce the possibility of disruptive behavior?" These steps must take into account the most common reasons or possible motives for misbehavior:

1. The student does not feel he/she belongs.
2. The student feels ignored or defeated.
3. The student is acting out the distress produced in a previous, unrelated situation.

If the teacher provides a nurturing atmosphere, then feelings of insecurity, loneliness, isolation, and anger can be tempered by the classroom atmosphere. The simplest way to do this is to share your time and attention with the students. While this sounds simplistic, the average industrial arts teacher might be surprised to realize that this basic form of human contact has gradually eroded. Ask yourself these questions with regard to the use of your time:

1. Where are you at break time?
   Teacher's lounge?
   Office?
   Working on a project?
2. How do you spend your time when students work in the lab?
   Correcting papers?
   Working with only a few students?
   Setting up for the next class?
3. Are you available to students who want to visit privately?
   Are you sought out?
   Do you listen or lecture?
   Have you taken professional courses in related fields of interpersonal relations?
4. Do you participate in extracurricular programs?
   Are you strictly a "shop teacher"?
5. Are you spending most of your instructional time in inefficient ways?
   Who supervises handing out materials or tools?
   Can a student take roll in your school?
   Can students learn to become accountable for cleanup?

It is easy to squeeze time from the day to catch up on planning and personal projects. It is equally easy to become absorbed in technical problems, hardware management, and faculty socials. Efficiently utilizing time to be able to attend to students and spread yourself around can pay large dividends in improving working conditions by improving classroom relations. Try to greet each student and find out what he or she is doing in school or in the community. This type of involvement is part of the 80 percent "positive feedback" that behaviorists claim is necessary to learning.

Case studies 1 through 6 are examples of behavioral approaches applied to typical discipline problems found in the classroom.

Case 1: Student Seeking Attention

This particular problem was described at the beginning of the chapter. The reader is also directed to chapter 3. Simply stated, the teacher can respond by:

1. Ignoring the disruptive behavior if it is minor.
2. Establishing in advance contingencies for such behavior.
3. Documenting the disruption and pursue the consequences.
4. Reinforcing the student when involved in positive behavior and make no reference to previous disruptions.

Case 2: Noncompliance

In this example the student refuses to cooperate, perhaps becoming involved in a shouting match or a similar power struggle. If the student is basically congenial and cooperative, then the teacher should explore the possibility that the student is expressing frustration with a class assignment. If this is the case, then reduce the assignment so that it falls within the student's ability. In this response successive approximations of the desired behavior are employed. For example, if a student refuses to wear safety glasses, instead of banning the student from the lab, allow participation where that student shows correct safety practices. Or, if a student refuses to clean up a work area thoroughly, begin by breaking down the cleanup job (task analysis) into components (e.g., clean the floor, clean the bench, put away tools, put away stool). The student earns points for each task completed. The percentage of the total points earned is the percentage of lab time earned (assuming it is desired by the student), with the remaining time spent in academic assignments.

If the student has already demonstrated an ability to do the task, then the problem may be more difficult. In this instance, a prior understanding by both the teacher and student regarding the response to noncompliance will potentially reduce the tension or nature of the power struggle that may ensue. A fairly typical sequence follows:

1. The student is warned verbally, unemotionally and without threat, and told the consequences for choosing correct versus incorrect behavior.
2. The student is removed from the class for five to ten minutes. This assumes that the student leaves voluntarily; otherwise, proceed to the next step.
3. The student is removed by an administrator for up to the entire period and must make up the time later. Depending upon the extent of the misbehavior, the student may not be readmitted the next day without making some restitution.

4. The student is sent home and a parent conference set up. This takes place when step 3 fails to produce results or if the behavior occurs more than once a week.

Case 3: Aggressive and Assaultive Behaviors

These types of behaviors are very serious and often result from deep-seated and complex personal problems if they occur with any degree of regularity. They present a very emotional issue.

Some specific techniques are presented first for the student who only occasionally becomes involved in a fight. These techniques can also become the general response to any classroom fight.

1. Require that the students resolve the conflict themselves without the instructor's attention. This can be done under supervision in the counseling office. Some instructors use "talk it out chairs." The students must sit facing one another and talk it out. In both cases they are not allowed back into class until they can assure the instructor that the conflict has been resolved.
2. Use special techniques. The United State Army requires two persons who have been fighting to wash the same window from opposite sides. They claim this injects humor, eases tensions, and allows combatants to work out a problem.
3. Intervene as referee. Students give both sides of their story before a teacher. The trick here is to make sure that each student presents his/her side without interruption. In 95 percent of the cases the fight was caused by miscommunication, which the instructor can help resolve. If these students later continue to fight, then step 1 or 2 may be more appropriate since the teacher's attention and interest may help to reinforce the fighting.

When a student has been warned but is unable to control his or her aggressive and assaultive behaviors, consider the appropriate degree of response. Two different levels of response to an outburst are *pre-mack* and *timeout*.

Pre-mack is the simplest and least intense method. In this instance two behaviors are related: the behavior the student wants (like lab time) and the behavior the teacher wants from the student (like talking quietly instead of yelling). The two behaviors are merely made contingent on each other. Thus, if the student talks quietly, then he or she can remain and have lab time. At other times it may be required that behavior in the special education room must be acceptable before shop class can be taken that day. Always state the contingencies in positive terms. Instead of saying "If you yell you cannot work here," you would say, "People who use a quiet, adult voice, can continue working in the lab with us."

Timeout is another correction technique. Here the teacher removes the student from the reinforcing environment. This is an *aversive* or punishing technique (assuming that the student wants to be a part of the class to start with). The traditional time out is usually removal from the work area where peers are involved. If the unruly student can have a view of the others working, the sense of exclusion is heightened. Time out should not last for more than ten minutes, after which the student is allowed to return and try again. Another modification is to restrict the student from participating in the full range of lab activities. For example, he/she would be restricted to tool disbursement or assigned a laboratory task that is different from that of the other students. If these methods fail, then a restricted placement in the program is suggested. Some options include:

1. Supervision from another adult (special educator or paraprofessional).
2. Working on the course in the special resource room until behavior improves.
3. Reduced time in the technology course with time increased contingent on good behavior (e.g., first and last five minutes of the day to hand out tools and collect them; later increased participation).

Case 4: Theft

Theft is a common annoyance most instructors face at one time or another. There are two basic types of responses:

1. Live with it and keep a low profile. A certain amount of theft is inevitable; it can be reduced with careful tool inventory and monitoring.
2. Make it the responsibility of the class to reduce theft. It becomes a class responsibility to replace or find the tool in lieu of losing lab privileges.

Some instructors argue the peer pressure endorsed in the second response can turn against the teacher; the class may feel apathetic toward or even supportive of the guilty party. The only realistic compromise is to discuss with the class *before* a theft occurs what the collective or group responsibility is and what are the consequences for the class when a tool is missing. The students often can make useful suggestions. Of course, a comprehensive and daily tool management system administered by the teacher and students can go a long way toward preventing theft, but this system requires a sacrifice of some class time.

Case 5: Unsafe Conduct

As with theft, education about safety can take place on a group level, although responses to infractions of the rules can be more personal. Industry and safety inspectors have established many different systems for safety management. Unsafe acts normally occur infrequently, but you must have some way of responding that shows that you have taken prudent action to eliminate the problem. First, it is wise to discuss with the class three categories of costs that result from accidents:

1. For industry, unsafe conduct results in loss of time, production, and personnel.
2. In the home, unsafe conduct results in injury or loss of property.
3. For individuals, this conduct could cause the loss of a limb, a job, and the respect of fellow workers.

Your response to one student's unsafe behavior could include the following:

1. Removal of student from the activity.
2. The requirement that the safety test be retaken or report be prepared before further work is allowed.
3. A statement sent home to notify the parents that the student is on probation.
4. Removal from the class.

Case 6: Problems that Are Difficult to Treat Individually

Often it seems impossible to pinpoint responsibilities for certain problems. An excellent example is a class that generally does not do an adequate job of cleaning up or that has trouble getting started on the day's work. Your first step is to observe. Try to determine the leaders of this undesired behavior, then record the frequency of the occurrence. The next step is to divide the class into teams or groups with the ringleaders in separate groups. Discuss the problem with the class and determine what needs to be done. Try to arrive at a reward suggested by the class that can be given at the successful conclusion of the program. The teacher then suggests the target behavior and goal based on the baseline. In this case it might be "Students will start work within three minutes of the lab session for five consecutive days." The reward is given to each group as it achieves the target. In this way an element of competition and group effort is injected. After all the groups have achieved the initial result, the reward is faded out. If one group is held back by an individual, then that person is put on an individual program so that the group can achieve the goal. In some situations it is not possible to fully achieve the goal. Therefore, once a baseline is established the goal becomes one of improvement. For example, if the ultimate goal were to reduce arguments or rude comments and the baseline were 10 comments per day, the initial goal might be reduction to 5 comments. If that is achieved then later the goal might become 0-1 comments per day.

CRITICISM OF BEHAVIORAL METHODS

Criticisms of behavioral approaches are not new. They are generally concerned with philosophical and moral issues, as well as with questions of practicality. These criticisms are presented here to provide a balanced picture. The theory and praxis of any method should be critically examined.

The questions that teachers typically ask about behavioral approaches are:

- How will students ever learn to operate without rewards?
- Won't students become dependent on highly specialized interpersonal relations?
- What about other students who are not being rewarded but who maintain good behavior?
- Won't students merely substitute one negative behavior for another since the cause for the behavior is not being treated?

There are many different possible responses to these questions. The purpose in raising these questions here is to stimulate thought, not to resolve these traditional criticisms.

Another significant question regarding the practicality of a reward/punishment system concerns its viability in a secondary school setting. In the middle schools behavioral methods are easier because the teacher seems to have more rewards at his or her disposal (free time, game time, field trips, and food, for example). The secondary teacher is working with an emerging adult with far broader needs and concerns. As a result, what is reinforcing to the adolescent is often something over which the instructor has no control. At this stage young adults are seeking peer status and membership, interaction with the opposite sex, job opportunities, and cars and stereos. Because this is so, the parents must become involved. It is they who usually dispense these favors or grant the privileges that allow students to gain access to these rewards.

However, it is also easy to "saturate" on behavioral techniques or have a "behavioral management burnout." This is typically caused when teachers and guardians attempt to tie all the things that are reinforcing to a student (privileges like car usage, allowances, Friday nights out) to a single behavioral contract. Sometimes the student responds with open rebellion to the entire authority matrix.

On the philosophical/moral side, behavior modifiers are often pictured as pandering to the lowest common denominator of educational goals: order and control. In a country based on the ideal of individual freedom but becoming increasingly dominated by technological methods of standardization, modular design, and the interchangeability of parts, we have a dynamic tension created by the needs of democracy and the needs of automated systems. Promoters of behavioralism are accused of translating engineering design principles into socially desired goals; of promoting a "mindless technician" who merely seeks to reinforce the status quo in the schools, rather than undertake a critical examination of the educational setting. Educational programming then becomes a simplistic pursuit of fitting the child to the environment while failing to examine the effect of that environment on the well-being of the student. Presented next is an alternative or supplementary approach to behavioralism. This alternate method is common to most teachers and relies on the dynamics of group discussions.

GROUP DISCUSSIONS

The use of group discussions in shaping and changing behaviors belongs to many different schools of psychology. The Adlerian model, which is part of the school of humanistic psychology, takes a decidedly nonbehavioral approach when using group discussions. This section presents a summary of the group discussion technique presented by Driekers, Grunwald, and Pepper in the their book *Maintaining Sanity in the Classroom.* Because it is readily adaptable by regular teachers, this technique does not require a great deal of professional preparation.

The main argument of humanistic psychology is that responsibility cannot be taught by coupling observable responsible behaviors with rewards. Rather, they assert that responsibility is a total experience that can only be acquired as the individual practices it and finds personal value or identity in being responsible. To accomplish this, Driekers defines a system of logical consequences in contrast to rewards and punishments. Group discussions are used to clarify values. The Adlerians stress democratic group processes over behavioral training. Today, many instructors use methods from both. That is, group discussions are used to augment individual programs, and democracy is practiced in consulting students about various concerns.

In the author's research of instructors who had been involved with mainstreaming in the technology curriculum (but who had no special education preparation), one teaching style continually surfaced. This was a democratic class atmosphere that demanded respect for students and teacher. The respect these instructors had for the students was apparent in the instructors' continual consultations with the students and in the solicitation of their students' advice, consent, and support for solving problems in the class. Likewise, they did not feel threatened when students proposed alternate methods and solutions. They involved the class in shared decision making. In addition, their involvement with the students did not end with the last bell of the school day, nor did they restrict their listening and counsel to subjects only belonging to technology or careers.

Logical Consequences

Driekers, Grunwald, and Pepper offer an interesting alternative to aversives and punishments with their concept of *logical consequences*. The concept asks that the student experience consequences for the actions rather than punishments unrelated to their actions. For example, if a student refuses to wear an apron while checking the battery fluid in a car, the logical consequence is possible acid burns on clothing. The instructor need not impose a time-out or some other aversive unrelated to the task. If a student will not follow directions on assembly of a project and ends with a piece of junk, then the product and grade become the logical consequence. Some consequences must be *applied*, and this is especially true where safety is a concern. In the case of the battery acid, students may be required to wear safety glasses, whereas they can take responsibility for wearing the apron. If a student does not wear the safety glasses, he or she will not be allowed to open the hood of the car. An instructor could establish the rule that anyone who breaks a safety rule cannot continue until he or she demonstrates the correct rule, fills out a form documenting his or her action, and enters into a probationary class placement for one week. This action is logical because it reflects the prudent action of teachers and protects other students.

Implementing Group Discussions

Group discussions have great application and importance in the secondary school because as emerging adults, the students want to have more say in what happens around them and to them. They may not be willing to accept a behavior modification program imposed by an authoritarian figure. They may simply not show up for class or remain unmotivated by the rewards you have at your disposal. When involved in a group discussion peer pressure is utilized to provide support and feedback. Peer acceptance seems to be at the center of the adolescent universe. Floy Pepper suggests group discussions are so important that time should be set aside each week for them.

It is important to consider carefully the role the teacher is to play in group discussions. The instructor is definitely not there to appear as guest lecturer or to control the group discussion to pursue his goals, or reflect his values. The teacher acts as the facilitator and guide. The teacher must feel secure in his or her relations with the students and be able to accept suggestions with which he or she does not personally agree. The instructor must be able to respect others' opinions and learn from them if he or she is to expect the same behavior from the students. As a facilitator, the teacher solicits the collective wisdom and support of the group. The facilitator must make sure that everyone has a chance to speak, that the group sticks to the topic, and that criticism is offered to others in a nonthreatening and positive manner. Later this leadershop role may be relinquished to other students.

If the teacher awaits a crisis before establishing a group discussion format, he or she may find the group will not coalesce under tension. Therefore, it is wise to introduce this discussion process early in the term around nonthreatening subjects. Probably the first group meeting will be used to set up the rules for discussion. They might include the following:

1. Taking turns to talk.
2. Private family matters are not to be discussed.
3. The time limit for the discussion.
4. How not to hurt other's feelings.
5. Possible future topics.
6. The reminder that these discussions are private and need not be repeated for mass consumption.

Often at the end of a group discussion it will be wise for the teacher to summarize what has been covered. The instructor may also ask the group to write down a specific plan to implement their ideas. In these beginning sessions the group might discuss a plan to carry out the first field trip. The discussion can focus on where to go, what rules of behavior should be established, when to go, and who should do what. The next round of discussions might be involved with the creation of some specific class rules, the design of a personnel system, guidelines for sharing tools, or ways to raise funds for materials. Start by telling the group *you* have a dilemma and need *their* advice. Students will respect this and realize that the instructor has limits around which he or she must operate. The final stage before approaching individual behaviors is to begin discussions about careers and technology. You could discuss futurism, science fiction, job markets, job experiences, recreational activities related to technology, the history of inventions, skilled versus unskilled labor, unions, and the like.

A lot has been accomplished at this point resulting in a gradual building of rapport within the group. Rules of operation have been established and practiced. The teacher has shown that he or she will seek advice and that everyone can work as a group on projects. This establishes the base for working together to help individuals. Now the group is seasoned and ready for helping with personal problems. The authors of *Maintaining Sanity in the Classroom* provide many insights into observing and interpreting individual behavior and personality; such terms as "family constellation," "recognition reflex," and "the four goals of misbehavior" appear in the text. While specific mastery of these techniques is not essential, it is useful when working on individual behaviors. When unsure as to how students might react, consult with the learning specialist who works with that student, or pursue training in techniques related to group discussions and counseling.

Let us illustrate the process by example. Jim refuses to do his part of cleanup each day. As a result the class takes longer to clean up before being dismissed. During the first round of discussions the instructor can introduce the problem at some distance by use of a parable. He might read a story about a worker in a steel mill who refused to put away his tools and how it resulted in an accident. Then he might ask the class the following questions.

1. Why did he feel this way?
2. Have you ever shirked your responsibility? How did you feel?
3. What might the foreman do?
4. Is it possible this person has no friends and is retaliating?

This gives the group a chance to begin to understand. The class begins to develop empathy as well as critical judgment that may serve them later in discussion. Jim might also get the message at this point. If he doesn't, it is important to bring the example in the next group discussion a little closer to home. This time the facilitator might discuss a group of events in the class that include Jim's behavior. One might ask if the group is satisfied with the personnel system, ask if the jobs are getting done, what changes could be made. In the next discussion the topic finally hones in on the problem if it has not disappeared. An introductory comment might be: "We have been discussing responsibility at work and our system of cleanup, tool dispersal, and so forth. I think today we need to examine how you as individuals have been operating. Jim, I noticed that you have not been cleaning up. Do you know why?" If he states his reason, then the group is off and running. If he doesn't know or want to say, then the teacher might proceed with, "Can I guess why? Is it because you think you are too good to clean up?" At this point Driekers' discussion on why students misbehave will serve as a useful resource in your guessing. At some point other students might also be allowed to guess. The dialogue that ensues will help Jim to see how others view his action. As this comes about and as some approximate reason for Jim's action begins to surface, it is important to find ways of helping Jim. At this point the teacher can turn to the group and ask them how they can help Jim to do his part. (Often by this point individuals have already offered such advice.) The instructor must inject into the discussion what the group likes about Jim because this teaches others how to offer criticism in sensitive ways. Formulate a plan that other students can partake in and have them monitor Jim's progress. In this case, the students might agree to a buddy system of helping Jim, or Jim may just agree to start cleaning up.

The author has found that the novice does not have to be a school counselor to carry out useful group discussions. (Even the experts do not have examples as clear-cut as the case of Jim.) However, it has been one of the most creative and exciting experiences for the author to see a group of adolescents strive to express themselves and help one another. It is even more profound when some of the students participating have themselves labored under the labels of "emotionally handicapped," "disadvantaged," or "retarded."

SERVING THE SEVERELY HANDICAPPED

The normalization principle should be applied to all learners. Integration of handicapped adolescents and nonhandicapped adolescents of the same age can assist both student groups to grow, develop, and accept one another. Many studies and projects have demonstrated that severely handicapped persons can learn skills relevant to industry (e.g., 16-step drill machine operations). Also, persons with severe disabling conditions can live and work in the community. A host of writers in the field of specialized training has cited many projects and studies that support the contention that severely handicapped citizens should be placed in the community to work and live because they are not only capable, but also it is less costly than institutionalization and more conducive to the training and development of these citizens. Normalization must be carried out within a continuum of community services (medical, counseling, therapy, respite, public education, etc.) and with a shift in traditional educational programs (academic/developmental) toward a curriculum that is life-centered (vocational education and community living).

This clearly indicates the important role technology education can play in this endeavor. While most technology instructors will infrequently have a student with severe impairments placed in their classes, they must recognize the value of a potential placement. Likewise, the student will be associated with educational specialists who will normally play a very active role in the placement and support for placement of their students in a regular class. These students may require that instruction be carried out in very precise or specialized modes in order to tap their unique way of learning. This section on serving the severely handicapped describes some instructional techniques and materials. However, working closely with the educational specialists will result in an individualized educational program that is best suited for student and instructor alike.

Placement Options and Curriculum Content

A logical place to start is with a pre-entry course set up for specialized training that will involve remediation and the teaching of basic skills. An example of this program is one the author initiated (case 1 in chapter 5). In this example students with moderate and severe disabilities were instructed in basic hand tools, operations, interpersonal skills, and work habits in a specialized class in a wood and electronics lab. Later, the transition was made to a regular two-week program in an electronics class that proved successful for everyone involved. The goal is always to move toward the least restrictive setting. The author further asserts that these students were very easy to instruct since they cooperated at all times. It was apparent to the author that these students were highly motivated and enjoyed the opportunity to participate in a hands-on environment in a technology setting.

Two examples of possible course content or methods are presented here. First, the reader may wish to consult the article "Strategies for Special Needs Teaching" (Bies 1979), which lists many general teaching goals. Examples of these are:

1. Looking, listening, and moving skills
2. Motor imitation skills
3. Task development skills
4. Self-help skills

Another model program is Project VOC at the University of Oregon in Eugene. They have developed through their Specialized Training Project methods for instruction and have applied them to school settings involving mainstreaming. Their program is based on "general case programming" methods applied to vocational instruction. They have task analyzed technology clusters and broken them down to prerequisite individual hand tool skills. The goals for training can be summarized as:

1. Teach the student to adapt to changes in work requirements.
2. Teach the student to respond to small discriminations demanded by work processes.
3. Develop the student's manipulation skills.

These goals can be approximately pictured as occurring in three phases:

Phase 1: Skill Acquisition
Phase 2: Applications
Phase 3: Work Habits

Phase 1 of this type of programming is oriented around skill acquisition. These skills come from standard lists of time-motion-study engineers. They are taught in isolation around a specific task and include the following:

Preposition-position (placing a screw in a hole)
Grasp-release (wire cutting)
Hold-delay-rest (soldering)
Assemble-disassemble
Transport
Use: Twist, squeeze, arc, drive, pull, push

These skills are taught on a difficult-to-easy continuum. First, just placing the screw is taught, then turning is added, tightening follows, and finally removal. Tools are introduced gradually: The screw is first turned by hand, then the use of large screws versus smaller ones, finally the screwdriver is used. To teach adaptive skills different size fasteners and tools are introduced. The manipulation is then done with the object in different positions. Finally, a rudimentary assembly could be added like nut, lock washer, and bolt.

We might call the next phase applications. Here the skill is generalized and adaptive efforts are reinforced. At this point real objects can be introduced. The local hardware store could supply you with knife switches, electrical outlet boxes, lawn sprinklers, and plumbing valves and faucets. In this stage sequence, alignment and working in a confined space are learned. Finer motor skills are required (usually finger dexterity). The student takes the objects apart and reassembles them first by using teaching aids and then independently.

The last phase would be on work habits. Here the learner's skill is already assumed. The student now works on work samples. Job lots are acquired, perhaps on a contract basis, and the student works repeatedly at the fabrication of one item. Stamina or length of work time is evaluated and strengthened. Rate is also examined as well as quality control.

Project VOC, along with the State of Oregon Mental Health Division MR/DD Program, has produced the *Resource Book Vocational Activities*. This is a source book of instructional ideas related to the types of pre-entry curricula being discussed. Figures 4-1 through 4-4 are examples of utilizing common items found in hardware stores that serve as the basis for teaching the skills in phases 1, 2, and 3.

Example of Instructional Technique

Individualized instructional materials that can be adapted to a student's individual pace and that utilize aides in the delivery of instruction can assist in the teaching of students with severe cognitive delays. If task analysis is combined with collection of data on student performance in a formated instructional program, one can attempt to apply behavioral methods to technical instructional goals. That is, by attempting to program the daily instruction in all its aspects—cues, reinforcements, punishments, methods of correction—a measure of control is exerted over the learning environment. Additionally, by documenting each student attempt, the instructor can begin to analyze and isolate barriers to instruction. Thus, behaviorally based instruction in this instance is merely the attempt to standardize and keep constant the multiple variables that go into any educational program so that only one variable at a time can be changed in order to determine its effectiveness. Additionally, this method helps to focus the learner's attention on the relevant skill at hand and not on extraneous events, changing formats of instruction, or idiosyncratic mannerisms of the teacher.

TITLE: Speedflex valve

Finger twist

PLACE OF PURCHASE: Plumbing section of a hardware store

MATERIALS: At least one speedflex valve (about $3.50 each)

Assemble sequence:

1 Task pieces

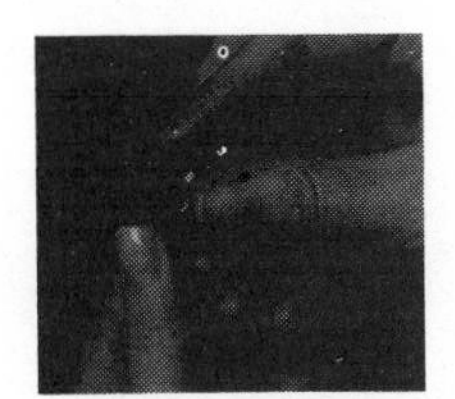

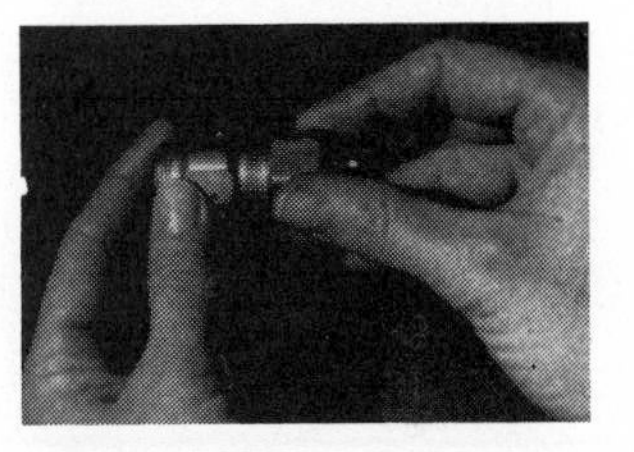

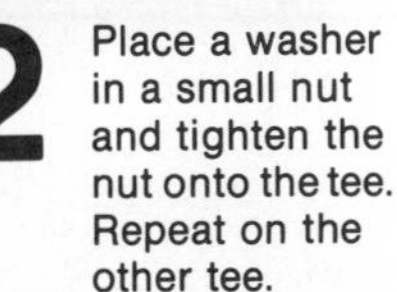

2 Place a washer in a small nut and tighten the nut onto the tee. Repeat on the other tee.

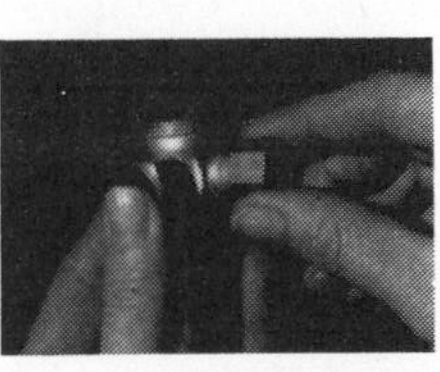

3 Place a washer in the large nut and tighten this nut onto the tee.

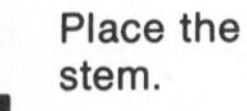

4 Place the stem.

5 Turn down the nut on the stem until the stem is secure. The task is complete.

Figure 4–1. *Finger twist. (Courtesy of Specialized Training Program, Center on Human Development, University of Oregon.)*

Pliers are used in many different ways. There are pliers that: **14**

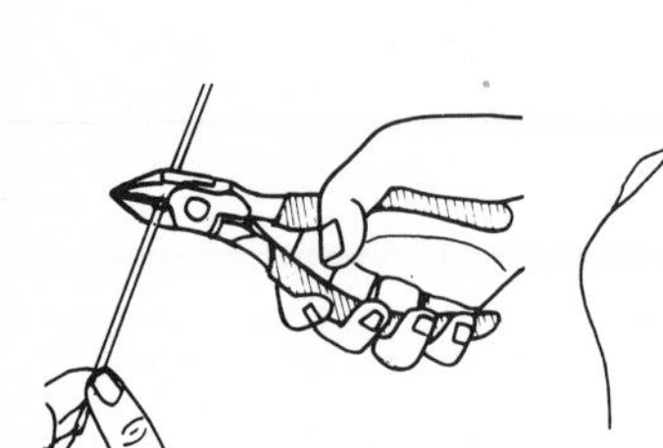

CUT

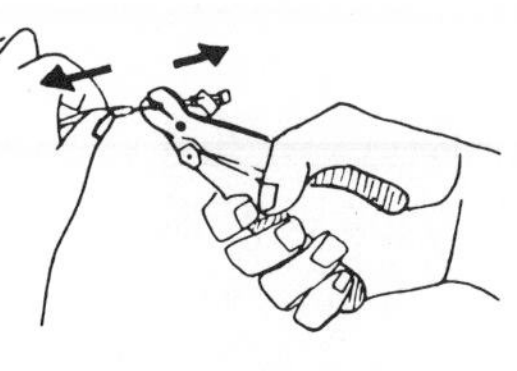

STRIP

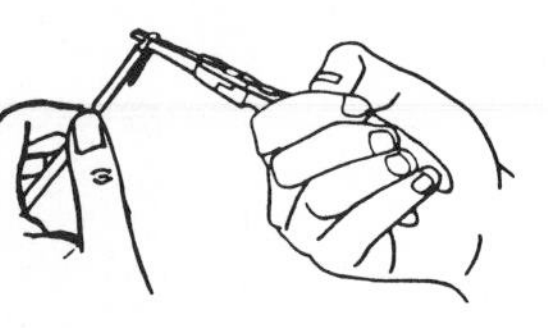

BEND

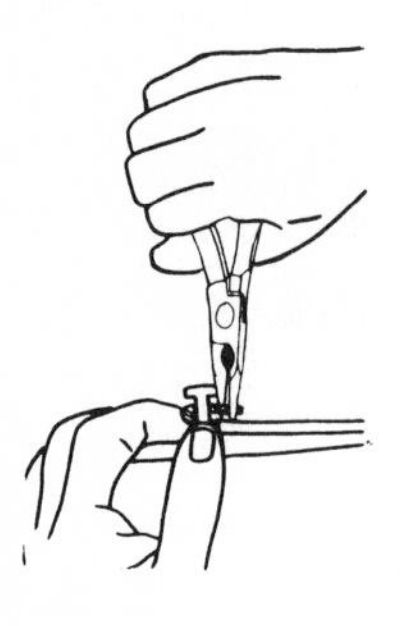

WRAP

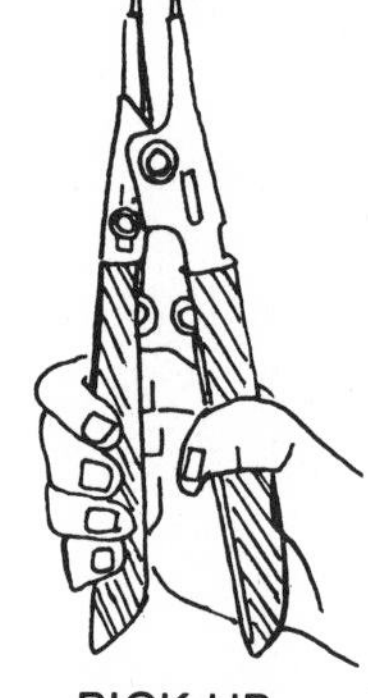

PICK UP

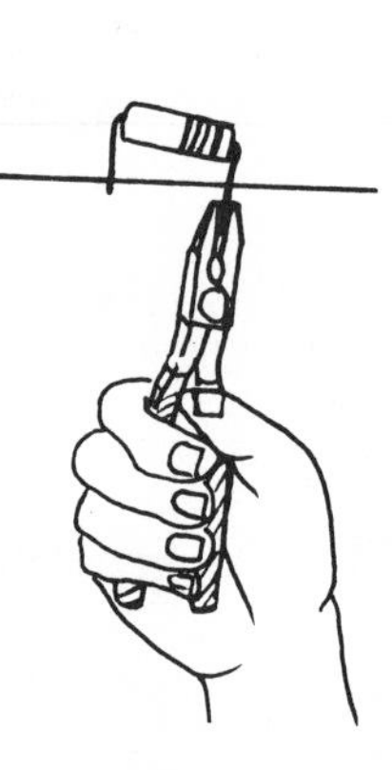

PULL

PLACE

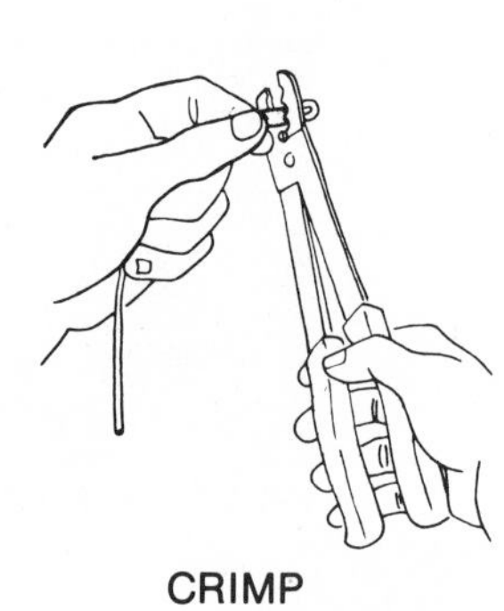

CRIMP

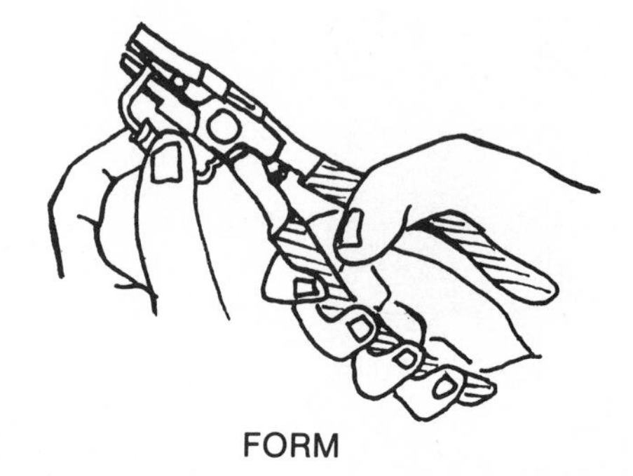

FORM

Figure 4–2. (Courtesy of University of Oregon, Eugene)

TITLE: Quick link and washers

Order

PLACE OF PURCHASE: Hardware store

MATERIALS: Quick link
Assented lock washer
internal lock washers
external lock washers
washers

Assembly Sequence:

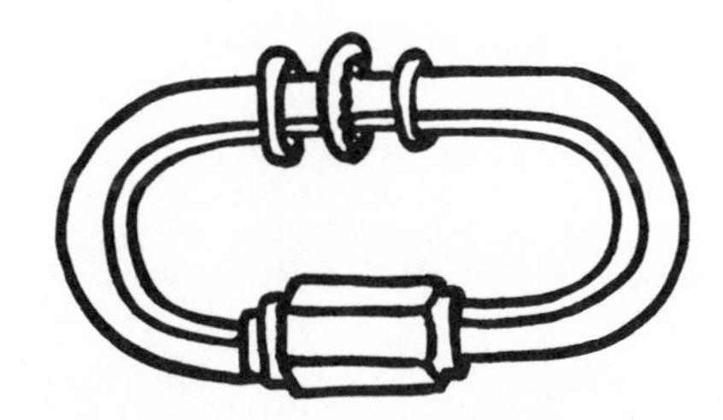

1 Prepare a model of what assembly *sequence* you would like to follow in placing washers on a quick link

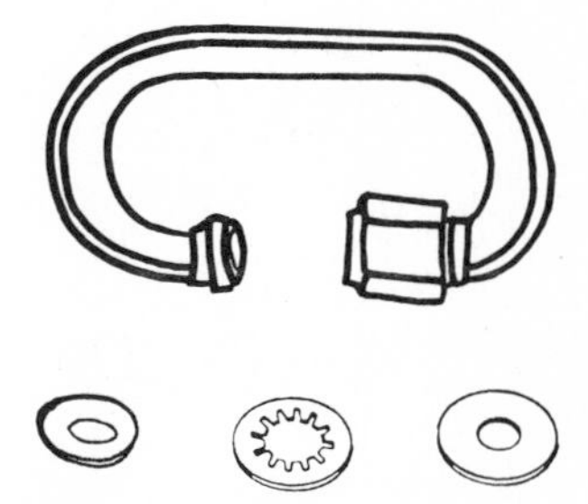

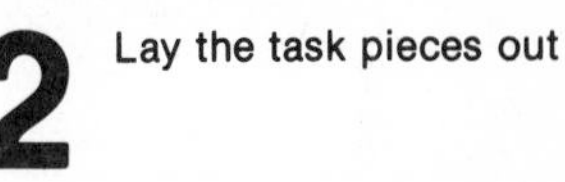

2 Lay the task pieces out

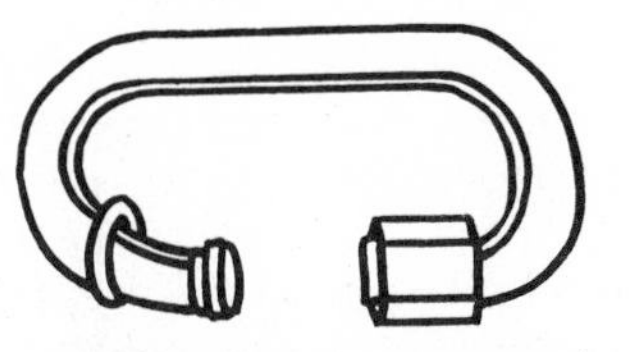

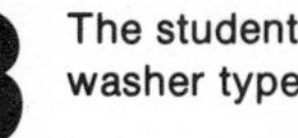

The student places the corresponding washer type onto the quick link . . .

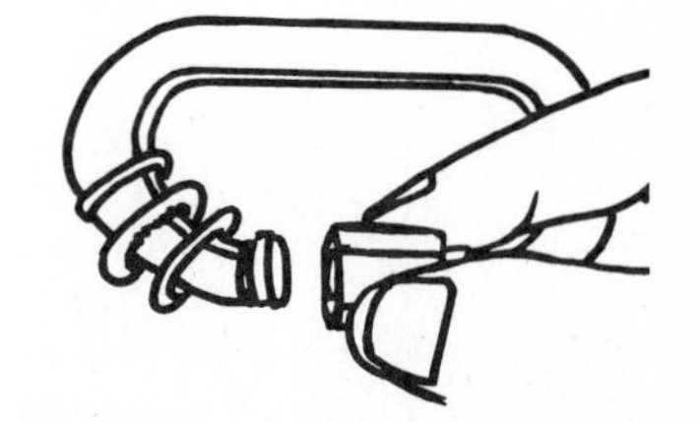

. . . in the order given on the model. Then close the link by turning it shut

▶ Note: You can use many combinations - you can prepare several different models and have a student make one of each.

Figure 4–3.

Package

TITLE: Packaging - Kits

PLACE OF PURCHASE: Any hardware store

MATERIALS: A plastic divider - 4 spaces
4 sets of screws - 10 screw size 4 x ½" phillips flat head
10 screw size 6 x ⅜" phillips flat head
6 screw size 10 x 1" phillips flat head
10 screw size 12 x ½" phillips flat head

Assembly Sequence: Xerox the picture model below and tape it on the lid of the kit.

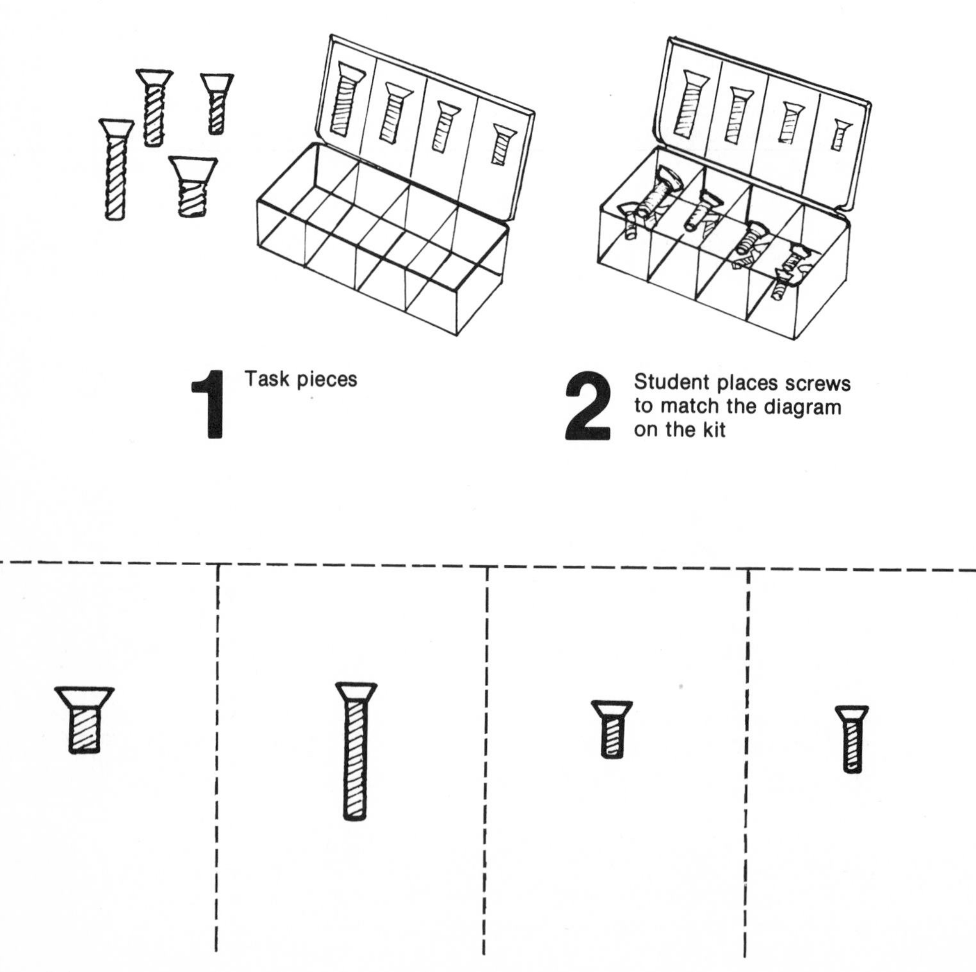
1
Task pieces
2
Student places screws to match the diagram on the kit

Figure 4–4.

SUGGESTED USES
Package
SCREW ANCHOR KIT
Instruction:
1 Kits can be packaged by *type* as with this screw anchor kit
BOLT KIT
2 Kits can be packaged by *size* as with this bolt kit
SOLDERLESS LUGS KIT
3 Kits can be packaged by a *distinguishing feature* as with this solderless terminal kit
BOLTS KIT
4 Kits can be packaged by *use* as with this wrench, nut, bolt and washer kit

Figure 4–4. (continued)

To illustrate this approach to learning, the example of instructing a student in a metals class is presented. The goal for the lesson will be to instruct the student in spot welding a sheet metal box for the purpose of assembly.

The beginning task analysis for this is presented as follows.

TASK ANALYSIS

TITLE: Spot-Welding a Box at the Flap
SAFETY: Face Shield, Gloves, Usage of Spot Welder

Steps	Prerequisite	Materials
1. Pick up box by inserting fingers into box with thumb on outside.	Two-hand dexterity	Box
2. Insert top edge between welder's tongs	"between"	
3. Slide to farthest right edge.	"right"	
4. Slide forward so that tong is centered about tab.	"center"	
5. Depress foot pedal control.	Use of foot	
6. Release foot after current flows.	Know when current stops	
7. Move box to left side.	"left"	
8. Repeat steps 4–6.		
9. Pull box back from tongs.		
10. Rotate box counterclockwise to opposite end.	"rotate, opposite	
11. Repeat steps 2–9.		
12. Go to storage area.	Location of storage area	
13. Stack back in bin on top of others unless all ready at top (start new stack).	"top"	

Note: Steps 8 and 11 will have to be listed in total on program.

An example of a formated approach toward delineating this task analysis is adapted from the methods used in the Vocational Careers Program at Portland State University. This is presented in Figure 4-5. In examining this approach it can be seen that everything for the instruction of the student is articulated in precise terms. This enables anyone to deliver the same approximate instruction daily, thus allowing the teacher to utilize student aides, peers, and paraprofessionals in the implementation of this instruction based on one-to-one student-teacher ratio. Likewise, this program can be used by other students and transported to other sites.

PROGRAM SHEET

STUDENT'S NAME ______________ SKILL AREA: METALS OBJECTIVE: Spot Weld box with 0 errors

DESIGNED BY: MK REVIEW DATE: ________ COMMENTS: ______________

MATERIALS: Face Shield, Metal box pattern Gloves, Spot Welder

Step Criteria:
3 consecutive attempts (move to next step)

Program Criteria:
10 boxes for 3 consecutive days with 0 errors

Step	Cue:	Student Response	R-reinforcement P-punisher	Schedule	Correction Procedure
1	"Pick up the box" point to box		R-social praise P"'No" . . . . (see correction)	1:1 after 2 correct attempts 1:3	"like this . . ." model correct technique then repeat cue
2	point to tongs	inserts box between tongs	same	same	same
3	"Start Here" point to corner tab	moves box to corner so that tongs are above tabs	same	same	same
4	"Center Tab"	centers tongs on corner tabs	same	same	same
5	"Weld"	student depresses foot pedal	R-touch student's shoulder P-"No"	same	same
6	Machine stops noise	student removes foot	R-social praise P-"No"	same	same

Figure 4–5.

In addition to specifying instruction, data must be gathered on each attempt of the student to follow each step. In this program a student is allowed three attempts at each step. If an error occurs, the precise method of correcting the student is described in the program. Also, a criterion for mastery of the entire unit is indicated. This guides the trainers in fading out this program. A date for reviewing the program serves to maintain the student's skill. An example of a data collection sheet keyed to this particular program is shown here. This data allows the instructor to analyze progress or lack of it. Because the individual instructional steps are so small, it is often difficult to ascertain the student's progress without collecting the data and displaying it in some form. There are many methods for displaying the data and a simple graph or similar approach can illustrate the progress being made over time. Special educators can suggest many different approaches for displaying data.

DATA SHEET

TRAINEE ____________________

PROGRAM TITLE: ____________________ AND NUMBER ____________

CODE: 1-Correct 0-Incorrect Ø-Correct needed prompt

(When incorrect, a 0 is placed; a correct response to correction procedure is not recorded; after the correction procedure and the cue is repeated, the new response is recorded.)

B-Baseline R-Review MC-meets criteria

	Date	**B**	**14**	**15**	**16**	**17**	**Review 12-4**
	Initial of trainer		MK	MK	MK		
	Duration of session		35 min.	30	40		
Step	**Step Descriptor**						
1.	Picks up box	001	011	111	111 mc		
2.	Inserts box	000	010	101	111	111 mc	
3.	Starts with right corner	000	000	001	101	111 mc	
4.	Centers on tab	000	001	011	111	111 mc	
5.	Depress foot pedal	001	111	111 mc			
6.	Removes foot	011	011	111 mc			

Comments:
Date/Name

14/MK Are ten boxes per day a realistic goal in a 45-minute period? We did three boxes.

17/MK Meets step criterion for program with *three* boxes.

The instructor will want to review the data sheet each time it is used. The purpose here is to diagnose problems encountered in learning or instruction. The following information may be useful in analyzing the data displayed in this data sheet.

1. Check to see that dates, names, times, and other information are filled out correctly. This will indicate the general accuracy and precision of the trainer; the information is needed to assess the trainer and the program.
2. If comments are encouraged, look these over and respond to them after examining the data.
3. Look for the following patterns to help analyze the data:
   000............ All 0s indicates that this step is not working.
   010,100,110 .... Inconsistent data might suggest that a new trainer be tried or that the original trainer be observed before changing the program.
   011,111 .... Indicates mastery of that step, depending on your step criterion.

Only attempt to change one thing in a program at a time in order to isolate the barrier or strength in the instructional program.

The first time a teacher looks at this form of instruction he or she is likely to ask, "How can I both instruct and take data?" This is a logical question. With practice and familiarity with the program it becomes a simple matter. However, the first time the teacher administers the instruction, he or she may ask another person to collect the data or the teacher may practice the program in advance until it is familiar. Likewise, the first time an instructor attempts instruction with these types of techniques, it is wise to work as a team with the educational specialist.

Rehabilitation Engineering

The field of rehabilitation engineering coordinates various concepts and techniques from a variety of different fields, such as engineering, medicine, systems information, and rehabilitation. Breakthroughs in medical and biomedical research will have increasing beneficial impact on the lives of severely handicapped citizens and involve such things as artificial limbs, orthopedic braces, surgery, medication, mobility aids, communication devices, facility engineering, and implantation of materials.

Project TALL in Missouri (discussed in chapter 5) has devised several methods for utilizing commonly found transducers for the control of office and other equipment by persons with severe physical disabilities. Other types of technologies employed include prosthetics (the replacement of body parts with artificial items), orthotics (used to correct physical defects), and neuromuscular control (functional electrical stimulation to bypass damaged nerves). Other devices include:

Powered chairs; modified hand controls for use of automobiles; lasers coupled to sensors for remote control of equipment.

Electromechanical systems to assist mobility and communication and control of equipment; sensors for eye or jaw movement that can be translated into commands over communication or other equipment for persons who are otherwise immobilized (quadriplegic)

Sensory aids for the deaf/blind:
- canes with electromagnetic or sonic sensors
- ultrasonic systems attached to glasses
- night vision aids
- optacon-tactile output device for use with reading
- reading machines using computers and laser scanners coupled to voice synthesizers
- vision prosthesis (direct electrical stimulation to visual cortex)

Biofeedback hardware used as an extension or augmentation of human systems:
- information feedback regarding prosthesis and the environment (heat detectors, pressure grasp, limb load monitors, step control monitors for correcting physical movement, sensors in mouth for speech correction)

Bionic ear implants

Apendix C contains a partial listing of rehabilitation engineering centers. Appendix D contains examples of efforts and products in this field.

EXPLORATIONS

Activities

View films by Marc Gold including "Try Another Way" (consult special education teacher).

Send for catalog of films from Wisconsin Vocational Studies Center, University of Wisconsin-Madison, 964 Educational Science Building, 1025 W. Johnson Street, Madison, WI 53706.

Consult bibliographies and organizations related to severely and profoundly handicapped (e.g., *Severely Handicapped Resource Guide*; Child Development and Mental Retardation Center, University of Washington, Seattle, WA 98195; and American Association for the Education of the Severely/Profoundly Handicapped, P.O. Box 15827, Seattle, WA 98115).

Visit a group home.

Devise a student practicum that mainstreams students in an industrial arts class.

Consult literature and films available from the Technical Assistance and Dissemination Network: Illinois Special Needs Populations, Turner Hall 202 C, Illinois State University, Normal, IL 61761.

Readings

In addition to consulting the references cited in the text, the following may be used to supplement information presented on behavioral teaching methods.

Bandura, Albert. *Principles of Behavior Modifications.* New York: Holt, Rinehart and Winston, 1969.

Krumboltz, J.D., and Thorensen, C.F., eds. *Behavioral Counseling: Cases and Techniques.* New York: Holt, Rinehart and Winston, 1979.

Strumphauzer, Jerome S., *Behavior Therapy with Delinquents.* Springfield, IL: Charles C. Thomas, 1973.

Thompson, Travis, and Grabowski, John, eds. *Behavior Modification of the Mentally Retarded.* 2d ed. New York: Oxford University Press, 1977.

Ullman, L. P., and Krasner, L. eds. *Case Studies in Behavior Modification.* New York: Holt, Rinehart and Winston, 1965.

REFERENCES

Bellamy, Thomas G.; Wilson, Darla; Adler, Ellen; and Clarke, James. *Specialized Training Program.* Eugene, Ore.: Center on Human Development, University of Oregon, n.d.

Bies, John D. "Strategies for Special Needs Teaching". *School Shop.* 38:19–20, May 1979.

Dreikurs, Rudolf; Grunwald, Bernice; and Pepper, Floy. *Maintaining Sanity in the Classroom: Illustrated Teaching Techniques.* New York: Harper & Row, 1971.

Edwards, Jean. *Vocational Careers Training Model: A Training Manual for Those Working with Moderately and Severely Handicapped Adolescents and Adults.* Portland, Ore: Portland State University, 1978.

Geiser, Robert L. "Behavior Mod and the Managed Society." *The Exceptional Parent.* 9:E2, Aug. 1979.

Rosenfield, Sylvia. "Introducing Behavior Modification Techniques to Teachers." *Exceptional Child.* 45:334–339, Feb. 1979.

Snell, Martha E., and Renzaglia, Adelle. "Characteristics of Services for the Severely Handicapped." *Education Unlimited.* 2:55–59 Feb. 1980.

State of Oregon Mental Health MR/DD. *Resource Book Vocational Activities.* Salem, Oregon, n.d.

The Urban Institute. *Report of the Comprehensive Service Needs Study.* Washington, D.C.: HEW contract 100-74-0309, June 23, 1975.

Wehman, P., and Hill, J., eds. *Vocational Training and Placement of Severely Disabled Persons.* Richmond, VA.: Virginia Commonwealth University, 1979.

Chapter Five

ACCESSIBILITY TO TECHNOLOGY EDUCATION

*I*n assembling the various elements presented in previous chapters, the barriers to mainstreaming come into complete focus. These barriers include stigma, social and curricular isolation, ignorance of disabling conditions, lack of planning or preparation for students, and lack of support for the regular teacher. The whole of this process results in de facto or institutionalized segregation. These barriers fall like a house of cards before professionals who blend their talents in the pursuit of an educational continuum of services designed for special needs learners.

By combining the salient points of the Life-Centered Technology Education model (see chapter 4) and student placement options (chapter 2), a community school can begin to develop such a continuum of services. This chapter describes specific instructional techniques, materials, and exemplary programs that help to anchor these models in the realities of the classroom. Entry-level, regular class, and post-secondary options are covered in turn.

Entry-level courses can serve a variety of learners. The instructional methods and materials presented can be directly used for learners with disabilities in the mild to moderate range, the range filled by the overwhelming majority of students labeled as handicapped. This core of material and methods can be adapted to the instructional needs of the students with moderate to severe disabilities by increasing the degree of individualization (greater task analysis breakdown, better student-teacher ratio, etc.). A caution must be borne in mind: the entry-level program should never be considered the sole prerequisite for placement in a regular class setting. Otherwise, these programs could become a holding pen or screening device. Every student needs to have access to integrated settings.

Regular class placements are discussed in light of teacher coordination and support options, which include peer tutoring systems and cooperative instructional agreements. In addition, instructional methods related to the adaptation of worksheets, safety instruction, and individualized instructional packages are included.

Finally, the chapter concludes with a description of possible postsecondary options. A supplementary appendix further includes examples of sources for instructional materials, curriculum materials, literature, and organizations that can provide additional resources and actual classroom or lab materials.

This chapter represents the consummate blending of two fields, the final picture in detailed focus. One must recall Hegel's admonition: "The sum is never equal to the parts." Likewise, one might at this stage recall a variation of Murphy's Law, which states that nothing is impossible for the person who only has to make suggestions. However, the program a teacher finally implements will not be nearly as complicated as the discussion of the elements or parts up to this point. Things have a way of falling into place, especially when this text is viewed as a resource book and not a cookbook. As the pieces to the mainstream puzzle naturally fit together, the instructor will surely find that the blending of two disciplines will be exciting, opening up new professional and personal vistas.

ENTRY-LEVEL CURRICULA

These curricula will focus on content and materials for programs that prepare students for regular class placements. Entry-level programs can be broken down into preentry and entry training classes. The former is more specialized and involves remediation or basic skill training. The latter can employ a variety of settings and staffing options: special educator, technology educator, team teaching, technology resource instructor, and others (see chapter 1).

The general goals for such a program are outlined below.

These initial goals are further broken down into numbered performance indicators that should be viewed as suggestions rather than as requirements.

I. To prepare the student to succeed in a regular technology class.
   A. To provide training in entry-level skills related to existing curricula.
      1. The student identifies different materials, tools, processes, and functions.
      2. The student uses measuring devices related to practical applications.
      3. The student performs written work similar to that required in the course (e.g., logs, worksheets, applications).
      4. The student uses course-related literature (e.g., catalogs, charts, operation sheets).
      5. The student interprets technical information (e.g., blueprints, models, plan sheets).

6. The student demonstrates study skills (e.g., attends to directions, demonstrations, note-taking, and reading assignments).

B. To provide training and experiences that encourage adaptive and age-appropriate behavior.
   1. The student masters the work habits and interpersonal skills needed in a technology setting: cooperation, cleanup, orderly work, safety, personnel system responsibilities.
   2. The student practices general safety: asks permission when required for equipment usage; reports accidents and injuries; asks for help when unsure; avoids horseplay.
   3. The student practices personal safety: rolls up sleeves; ties back hair; removes jewelry; wears safety apparel.
   4. The student practices appropriate interpersonal skills: acceptable grooming, appearance, and hygiene; conversational skills; assertiveness; ability to handle teasing; ability to accept criticism; ability to work independently; ability to concentrate on work.

II. To promote a self-concept that supports participation in our technological society.
   A. To explain the role of technology in society.
      1. The student identifies careers related to technology.
      2. The student identifies problems associated with technology.
      3. The student identifies current events associated with technology or with the history of technology.
      4. The student identifies consumer skills related to technology.
      5. The student identifies avocational applications of materials and processes.
      6. The student identifies nonvocational applications of technology (e.g., health care, rehabilitation, community use, access, communication).
      7. The student identifies personal interests and aptitudes related to technology.
   B. To reinforce the student's efforts and provide individualized instruction and evaluation.
      1. The student can repeat experiences or the course for greater mastery of skills.
      2. The student is evaluated on improvement, not on comparison with other students.
      3. The student can work at his/her own pace; instructional learning packages can be utilized.
      4. The student receives help with entry-level skills as needed.
      5. The student's efforts in skill mastery and personal adjustment are acknowledged.
      6. The student's instructional goals and instruction are kept within his/her ability range.

7. The student receives reinforcement for efforts outside class (e.g., awards, letters).
8. The student is asked to contribute his/her skills by helping others.
9. The student leaves class with a useful skill or product.
10. The student assesses the course.

III. To provide a smooth transition between specialized courses and regular courses.

A. To provide experiences in the regular class.
1. The student views films related to or included in the regular technology class.
2. The student participates in class try-outs and short-term mini-units.
3. The student participates as a teacher's aid (e.g., with tool dispersement).
4. The student attends open house and field trips with the regular class.

B. To involve the regular course instructor.
1. The instructor attends the IEP conference.
2. The instructor visits the entry-level class and describes the technology curriculum.
3. The instructor establishes a cooperative instructional agreement with the referring teacher.

This outline partially indicates the gap in educational services that can be bridged within technology education. This gap occurs when students with developmental delays in basic skills or behaviors are left segregated or else are mainstreamed without preparation and support. The advantages of providing an entry-level program can be realized by a variety of learners for whom the regular class offerings appear inadequate. This is not an attempt to lump different groups of students under a common rubric of "special needs"; it is merely an attempt to expand the utilization of special resources and thus broaden its rationale. An example of this is the ZVI Woods Technology program discussed later in this chapter. The materials were designed for students with limited reading ability and the materials were used by a variety of learners. The creator of the program claimed that by using his picture-based instructional packages all learners benefited.

Case Studies

Three case studies are presented here to illustrate how one might begin to implement an entry-level program. These examples do not constitute a comprehensive program and are based on the author's personal experiences, before he received formal training in special education. These initial experiences, combined with later instruction in special methods, provided the basic framework for the conceptualization of a Life-Centered Technology Education model.

Case 1

Goal	To prepare eight to ten students from a self-contained classroom to participate in a technology laboratory (electronics or woods). Students were in a program for trainable mentally retarded students (e.g., developmentally delayed with moderate to severe handicaps). This was a one semester, transitional program.
Method	A team comprised of career education paraprofessional, special education, and technology education instructors secured time and space in the "lab" (during the teachers' preparation period). A program was designed that would be team taught. It addressed technical skills, safety instruction, and work habits. These skills were coordinated with a safety instruction unit in the student's homeroom.
Content	
Woodworking Skills	Nail identification, hammering, nail removal, gluing, use of drill press and jig saw and related safety.
Classroom Skills	Safety instruction, use of drawings, measuring, group discussions and role playing related to the instruction of appropriate personal behaviors.

Typical Instructional Activities

Instruction was based primarily on individual needs, and not prerequisite vocational curriculum needs.

Safety

Classroom

A safety chart was studied in a group and each student had to identify the concepts illustrated on the chart. In addition, the students took turns leading the group discussion, which gave them a chance to practice related communication skills as well as reinforce the lesson content. Typical instruction was related to:

- personal safety (tie hair back, roll up sleeves, remove jewelry, use eye protection)
- communication skills (how to ask for help, the need for asking permission before operating equipment, eye contact)
- work habits (consequences of horseplay).

Lab

The above concepts and rules were reinforced in the lab setting.

Work Habits

Classroom

Interpersonal skills were discussed in a group setting. These included such things as:

- response to teasing
- communication skills (eye contact, greetings)
- hygiene and dress
- taking breaks and conversation skills
- responding to humor and jokes

Lab	Role playing situations were set up in the lab to test and reinforce the instruction received in the classroom. Some of the following formats were used: • student was given technical instruction in setting up an ohmmeter and encouraged to ask for further assistance or help • in a group discussion, students discussed responses to teasing, then a peer instructor simulated a situation with teasing • bells were rung for breaks and informal conversation was initiated • students had to identify which of two workers was not too busy to be asked to help the student • a personnel system was set up for tool dispersal and collection, involving all students.
Technical Skills	
Classroom	Some students began work on identifying and matching pictures with objects while others worked on rudimentary measuring principles.
Lab	Students received instruction in entry-level skills or assistance by utilizing three instructors and two student aids. Some students did not need this instruction and could work with minimal supervision.

Transition to Regular Class (1)

At this point students had mastered the basic skills that could lead to their effective participation in a regular class. Likewise the instructor had become more familiar with the students and thus more confident, which gave impetus to pursuing the next logical step: transition to regular classes. During the regular electronics class there occurred a two-week period during which students worked on personal projects. It was during this period that students from the special education class were invited to come in for a mini-lesson. Prior to the entry the regular class held discussions regarding disabilities, stigma, personal experiences, and the rights of handicapped citizens. At the end of these discussions half of the class volunteered to act as peer instructors. A system was set up for guiding, grading, and orienting the peer instructional team. The author served as the general coordinator with the special education and career instructor taking an active role in guiding the peer instructors. During the mini-lesson peer instructors taught the special needs peers some of the skills that had been previously taught in electronics: wire stripping, soldering, house wiring switches, and repair of cord caps (plugs). Once a week a "staffing" was held with the peer instructors to evaluate the progress to date and to share experiences. Peer instructors took data on their student's performance and this data collection served as the basis for granting credit to the peer tutors. An example of the format used is illustrated in Figure 5-1. Whenever a student received a 0 the peer instructor had to comment why, and whenever a student received three consecutive 1s the step was considered mastered. Review as well

TRAINING SHEET

Student's Name ______________________ Peer Instructor ______________________

Directions: Peer instructors are to record your daily instruction and progress on this paper. It will be handed in at the end of every week. This will be used to grade the instructors and plan the student's next lesson.

Code: 0-Cannot do without assistance. Three consecutive 0s, see teacher
1-Can do independently. Three consecutive 1s means mastery, and skill need not be reviewed

Skills	Days/Initials					Comments
Work Habits/Interpersonal						
Greeting						
Eye contact						
Adult behavior						
—describe if incorrect:						
Quits work on time						
Cleans up						
Asks for help when unsure						
Safety						
Avoids horseplay						
Sleeves rolled up						
Jewelry removed						
Eye protection worn						
Asks permission						
Review of these rules						
Technical Skills						
Names tools						
Demonstrates tool safety						
Operations:						
Strips house wire						
Strips component wire						
Makes solder joint						
Solders single strand						
Solders multi-strand						
De-solders components						
Solders printed circuit board						
Review of steps						
Student Evaluation of Peer Instructor						
+ . . . good - . . . poor						
? . . . not sure						

(On back write down any additional comments, observations or suggestions)

Figure 5–1.

as evaluation of the peer instructor can be built into this format. Many students who had themselves relied on tutors showed new enthusiasm and ability upon becoming peer instructors. The students were very enthusiastic at the end of the mini-lesson and asked that it be continued. While this could not be accommodated, the next type of transition was implemented.

Transition to Regular Class (2)

Students were allowed to return to the regular class but this time on an individual basis. The special education teacher kept some of the tools and materials in her room so that students could continue to work with them. She selected the students whom she thought would benefit the most from a regular class placement. This resulted in one student who was enrolled full time in the beginning or exploratory I.A. drafting class. Another student worked with advanced electronics students who had time available to work with the student with the special need. At other times, this student pursued an individual project that stressed good work habits.

Case 2

Goal	To create an entry-level exploratory program for students in a self-contained class for emotionally handicapped learners. This was part of a six-week student practicum, and as such, only the initial structure and stages for implementation were established. The goal was to offer simplified work projects for students who were apathetic and lacked practical or successful experiences in the technology curriculum.
Method	Work stations were set up in the resource room. The content came from those regular technology teachers who seemed most able to accommodate the special needs students (e.g., electronics, drafting, metals). Instructors donated materials and advice. The program was tied to the existing behavior management system in the special education room. This meant that students had to earn the privilege of free time in order to participate in this program. Earning this privilege was based on positive behaviors in the room. The students could also use work on these projects to supplement a point system based on required academics.
Format	Materials were stored on book shelves. As a student earned free time he or she received individual assistance at a work table. Additionally, once a week group discussions were held with topics related to work environments and personal behaviors or outlooks. Finally, the student was offered the option of visiting a class that related to the work samples.
Content	See Figure 5-2 on page 154.

Typical Instructional Activities

Within six weeks more than half the students were able to try out one or more project samples. One student who was classified as severely emotionally handicapped was able to master the initial setup of an ohmmeter. This student was later taught basic principles of behavior modification (e.g., cues, praise). Following this, he offered to serve as a peer instructor in a program serving students designated as trainable mentally retarded. This proved stimulating to the students and peer instructor who experienced a renewed

Student ________________________________ Week of __________

Directions:
In your free time examine the materials and tools on the bookshelf.
Choose a project you might like to make.
Ask the instructor for a demonstration or help as you need it.
You can earn two points per completed step; these can be added to your weekly total. To receive credit, the teacher must check you off when you have completed each step.

Work Habits
1. Each day you will be graded on your work habits.
2. Failure to earn 80% on this score will mean losing the privilege of working on your project.
3. You can earn two points for getting 100% on your work habits.
4. Before you leave, have these items below checked off (forgetting will mean a "-")

Cleaned Up					
Inventoried Tools					
Shared Materials					
Helped Others					

Electronics

____Strip component or house wire
____Solder wire joints
____De-solder five components from board
____Set up VTVM (meter)
____Test a SPST switch
____Read a resistor with meter
____Read color code
____Wire kit from pictorial
____Wire kit from schematic
____Memorize eight components symbols

Metals

Embossing

____Develop design on paper
____Emboss design on metal
____Apply finishes in metals lab

Ring

____Develop design and measure finger
____Cut out metal with saw
____Cut metal to size and remove burrs

Welding

____Identify tools from pictures or arc/gas welding
____Take a written safety test for arc or gas
____Lay a bead in the welding lab

Drafting

Sketching

____Sketch horizontal, vertical, and circular lines
____Sketch an invention of a machine from the future
____Sketch floor plan for your house

Geometrics

____Make a design using drafting instruments
____Bisect an arc, line, and angle
____Attempt geometric problem of student's choice

Orthographic

____Make three drawings of your choice (60% correct)

Architecture

____Make a house floor plan using drafting instruments
____Make a cardboard model of house

Weekly Score
Work Habits

Projects	
	Total

Figure 5-2. Industrial education explorations

Group discussions were held with the following typical points of discussion:

1. What is the best way to handle criticism?
2. How do you cope with teasing?
3. What do you want to do when you leave high school?
4. What are entry jobs typically like? (e.g., routine, boring, low pay)
5. How can one cope with boredom and stress?

Transition to Regular Class

For most students transition involved a visit to a related class (drafting, metals, etc.). For others, it meant attempting to salvage an existing placement in a regular class by reducing the student's participation to assisting in the tool room until behavior improved. For yet other students it meant discussing with them options for future classes that involved the technology curriculum.

Case 3

Goal — To provide a student with the opportunity to explore a vocational reprographics class. In this instance the student was only mildly handicapped. The purpose was to allow the student to acquire an awareness of the technology so he and the instructor could determine if future enrollment would be warranted. It also allowed the student the opportunity to learn some beginning skills should he later decide to enroll.

Method — The teacher in a career education class identified a student with an interest in graphics (e.g., wanted to work with machinery, had careful and precise work habits and skills). A cooperative arrangement was set up between the reprographics instructor and the special education instructor to carry out this vocational class exploration.

Format — The student was placed in the advanced reprographics class under the tutelage of an advanced student. The visiting student's performance in work habits and skills were daily recorded and assessed by the peer instructor on the form used in Figure 5–3.

Content — The student worked on the production of a note pad. During the course of this the student was able to observe and participate in the following skills:

- basic layout preparation (paste-up and hand composition)
- measurement techniques (copyfitting)
- reproduction photography
- basic stripping and layout of a flat
- preparation of an offset printing plate
- offset fundamentals
- trimming and binding finished product

OCCUPATIONAL EXPLORATION: GRAPHICS

Student ______________________ Date __________

Peer Instructor ______________________

Week of __________ to __________

Directions: Peer instructor grades student and then student grades the peer instructor. Please add comments when score is above or below a 2. If the student or peer instructor is unsure of what to do, be sure to ask!

CODE

Safety and Work Habits

1. . . Poor/Not Done
2. . . Could be improved
3. . . Good, Consistent

Specific Vocational Skills

1. . . Student has observed skill
2. . . Student was assisted/practiced
3. . . Student was independent in the skill

SKILL AREA	DAILY SCORE					COMMENTS
	M	T	W	TH	F	
SAFETY AND WORK HABITS						
1. Ties back loose clothing						
2. Exercises caution around light tables and machinery						
3. Mature behavior, NO horseplay, no experimenting						
4. ASKS WHEN UNSURE						
5. Other:						
6. Other:						
7. Attendance: 2 points-on time 3 points-present						
8. Appearance: safe, appropriate, Hygiene: clean, neat, acceptable						
9. Accepts criticism (asks questions, tries to change)						
10. Corrects errors						
11. Gets along with others						
12. Concentrates, steady work output						
13. Follows directions						
14. Show initiative: eager, seeks new work, eager to learn						
SPECIFIC VOCATIONAL SKILLS (list, add others to back as needed)						
15.						
16.						
17.						
18.						
19.						
20.						
STUDENT EVALUATION OF PEER INSTRUCTOR Use "Safety and Work Habits" Scale at top of page.						
A. Instructor was friendly						
B. Instructor was patient						
C. Instructor was mature/adult						
D. Instructor was knowledgeable						

(on back put questions, suggestions, other comments)

Figure 5–3. Vocational class exploration: peer tutoring checklist

Typical Instructional Activities

Generally, the visiting student observed the advanced student perform the tasks and then later was asked to help. The visiting student got to practice several of the skills on his own. The experience took a little over two weeks and the checklist in Figure 5-3 served as an alternative source of credit in the special education career class. During this time the special education teacher did some follow-up of a minor nature.

Transition to Regular Class

The student's experiences were successful. He decided to enroll in the class for the following semester.

Summary

Once the instructor feels confident and achieves some initial success in working with special needs learners, it does not take long for new ideas and variations of established technique to blossom and suggest expansion or alternatives to current class offerings. This will logically lead to an entry-level program that becomes a formal class offering within the larger technology curriculum. These sorts of experiments will have ups and downs, not unlike other teaching assignments. However, the larger view of the effort will surely be measured in termes of success for all involved.

Creating Entry-Level Instructional Curriculum

An entry-level program can serve as a vital link to mainstreaming if it has the following qualities:

- A coherent and comprehensive structure
- Is adaptable to individual student needs and abilities
- Teaches skills that will assist the student in the regular class

This curriculum cannot be developed in a vacuum but must involve special educators and technology educators in an active partnership. The following three phases or stages can act as a general guide in systematically working out the curriculum:

1. Analyze existing courses and inventory the barriers to mainstreaming.
2. Summarize prerequisite skills that lead to regular class participation.
3. Create instructional materials that can be easily monitored for individual student performance.

TECHNOLOGY CLASS CURRICULUM INTERVIEW QUESTIONNAIRE

Course:____________________ Date__________________

Interviewer: __________________________ Instructor:__________________________

1. Go over entire curriculum from beginning to advanced courses and list courses related to this technology.
2. List units of study from class syllabus for exploratory class.
3. Describe instructional techniques used and their percentage of the total instruction:
   ____lecture
   ____demonstration
   ____text reading
   ____written work
   ____lab work
4. List present learning options within course:
   ____individualized learning stations/instructional materials/ rate of learning
   ____individualized testing, use of competencies, credits
   ____individualized projects
   ____alternative testing
   ____1:1 support, buddy system
   ____other
5. List most common difficulties for average learner in class.
6. List most common difficulties for slower student in class.
7. Describe previous experiences with handicapped learners, mainstreaming, or teacher contacts.
8. Describe how the following learner deficits could be accommodated:
   ____inability to read
   ____inability to read technical literature
   ____limited writing ability
   ____little math or measuring ability
   ____inability to perform basic computations
   ____inability to use fractions or decimals
   ____undeveloped fine motor coordination
   ____inability to follow lengthy directions, poor retention
   ____inability to stay on task independently for more than 30 minutes
   ____immaturity
9. Discuss ability to take part of course or utilize existing materials for specialized training or placement.
10. Discuss useful forms of support for the technology instructor.
11. Investigate accessibility to machinery, work places, storage, bathrooms.
12. Discuss recommended traits, behavior, related interests for the course.

Figure 5–4. Technology class curriculum: Interview Questionnaire

Curriculum Analysis and Development

The best approach to analyzing existing curriculum options is to visit the teacher and the setting where the instruction takes place. The initial visit can be based around the interview questionnaire in figure 5–4. The questionnaire begins with an inventory of the course content and proceeds to pinpoint areas of difficulty in the regular curriculum that special needs learners may encounter. The following steps can then be used in the development of curricular materials for entry-level classes.

1. Summarize the findings from the initial curriculum analysis and select instructional materials that reflect the analysis. Two methods to consider are:
   - Identify those initial hands-on and academic skills that are critical to success in the class. An example of a resultant product is found in figure 5–5. This is a task analysis for a mini-course that is part of the prevocational program in the North Clackamas School District in Oregon. It is based on the entry-level skills needed for participation in the graphics cluster classes at the local vocational high school.
   - Group skills from the course based on common learning or instructional modes, and then find hands-on and academic projects which incorporate these skills. Figure 5–6 is a summary of prerequisite skills needed in a welding class grouped by common skill types: academic skills, physical aptitudes, personal traits, work habits, and specific vocational skills. This summary sheet became the basis for the design of a welding work sample used in a vocational evaluation program in the Thurston County Cooperative Special Services in Washington.
2. Determine how the student is to be evaluated. Several approaches can be used, two of which can be seen in the preceding examples:
   - The skill mastery level attained by the student after training is embodied in figure 5–5 where the student's work quality is recorded.
   - Evaluate the degree of support the student might need in the regular class. This approach is reflected in figure 5–6, which tends more toward evaluation.

   Both approaches look at similar student behaviors with slightly different ways of presenting the findings.

Often many instructional materials are assembled into a larger battery of training materials reflecting an entire cluster or the entire technology curriculum in the school. The specific types of instructional material and format used will determine how instruction is delivered, how data are gathered and summarized, and what staffing is required for the system.

Another approach to the development of curriculum is the adaptation of existing courses at the exploratory level. The simplest modification is to pair special and regular educators in team teaching situations. An example of the application of previous principles toward team teaching is found in the Madison Metropolitan School District in Wisconsin. The curriculum is embodied in their Basic Class program. The following is the course syllabus.

THE BASIC CLASS

Course Syllabus for the Basic Class Curriculum. Madison Metropolitan School District, Madison, Wisconsin. The entire curriculum includes agriculture, business education, home economics, art, and industrial arts.

Basic Art

Projects	*Duration*
Introduction	1 Week
Perspective	4 Weeks
One-point	
Two-point	
Barn Drawing and Coloring	
Color Chart	2 Weeks
Papier-Mâché (six projects)	3 Weeks
Portrait Drawing	6 Weeks
Face Parts	
Portraits of Other People	
Famous Person	
Seasonal Art	1 Week
Op-Art (six drawings)	3 Weeks
Painting on Canvas (six paintings)	9 Weeks
Introduction (How to)	
Drawing and Getting Ideas	
Building and Starting to Paint	
Work on Paintings	
Framing	
Clay	4 Weeks
Figure Drawing	3 Weeks
Watercolor (six paintings)	2 Weeks

Basic Industrial Arts

Projects	*Duration*
Introduction and Orientation	1 Week
Maps and Symbols	1 Week
Measurement	1 Week
Drawing Interpretation and Introduction	2 Weeks
Shop Safety and Orientation Sheet Metal Project #1	2 Weeks
Project Drawing: Tool Tote	2 Weeks
Shop Tools and Safety	1 Week
Sheet Metal Project #2	3 Weeks
Welding Introduction and Shop Safety	1 Week
Welding Projects	5 Weeks
Foundry Introduction and Shop Safety	1 Week
Foundry Projects	4 Weeks
Woods and Metal Project Introduction	1 Week
Woods and Metal Project	4 Weeks
Graphic Arts Introduction	1 Week
Letter Press	4 Weeks
Silk-Screening	4 Weeks

This program began in 1974 and is described as follows:

- The Basic Class curriculum is developed around the skills students need in order to succeed in the regular vocational program.
- The vocational education teacher is responsible for providing instruction and making appropriate changes in learning activities

GRAPHIC ARTS MINI-COURSE

Class Syllabus (Task Analysis)

Day 1:
Compose headlines using letters and composing stick.
Copy production (using nine sentences), produce a line drawing, make a self-portrait (using camera).

Day 2:
Make lithographic negatives (using portable camera and processor).
Make prints from negatives.

Day 3:
Strip negatives on a flat and correct with opaque.

Day 4:
Use plateburner to burn a plate.
Develop plate with proper chemicals.
Print the plate on an offset press.

Layout Evaluation Form

Number ____________

Layout Evaluation	Good	Acceptable	Not Acceptable
Content Headline Copy			
Design General Illustration			
Acceptability Printing Appearance			

Evaluator ______________________________ Date ______________

Figure 5–5. Graphic arts mini-course (Courtesy of Richard Michealis, Special Education Instructor, North Clackamas (Oregon) School District.)

and teaching techniques as they relate to the student's learning style.

- The special education teacher is to help the vocational teacher identify the exceptional education student needs and learning style and to provide assistance in the modification of student learning activities and teaching techniques.

SUMMARY OF WELDING CLASS PREREQUISITES

Directions: Summarize the student's performance on this prerequisite checklist. Then make a summary recommendation regarding possible placement in a welding class. Use the following code (some parts may require interpretation of time, errors, or observation.)

4. . . Work is acceptable in quality/time; student is independent.
3. . . Work has some errors; student needs more training and requires supervision.
2. . . Student can only perform task with supervision and further training.
1. . . Currently student shows no interest or ability in task; task not completed; training not recommended.

Interpreting scores:
4. . . . Regular class placement
3. . . . Support is suggested
2. . . . Adapted or modified placement recommended
1. . . . Placement in class not recommended

Skills	**Student Performance**			
Academic Skills	1	2	3	4
Reads technical literature				
Reads safety literature				
Takes notes				
Answers worksheets				
Measures accurately to ⅛"				
Reads and uses fractions and decimals on plans				
Interprets charts				
Physical Aptitudes				
Good near vision				
Good fine motor coordination				
Ability to crouch, kneel, crawl				
Stamina/endurance				
Personal Traits/Work Habits				
Tolerates fumes, noise, dirt, heat (sparks)				
Wears protective clothing & equipment				
Ability to handle frustration/patience				
Able to listen and accept criticism				
Desire to improve self				
Interest in taking related courses				
Cleans up				
Safe attitude/organized person				
Able to work alone (in booth)				
Specific Vocational Skills				
Ability to inspect errors				
Ability to use hand tools and do assembly work				
Ability to use hands together (bilateral coordination)				
Ability to run a bead (optional)				

Summarize Recommendations:

Figure 5-6. Summary of welding class prerequisites.

Design of Entry-Level Instructional Materials

After articulating the goals and analyzing the curriculum for an entry-level class, it is time to develop the specific instructional materials. Three different types of instructional materials and methods are described here: individual self-administered instructional units; small group exploratory instructional modules; and adapted exploratory class projects. While all three result in slightly different staffing patterns, they share common design principles. Whether the instruction takes place in a technology lab or a special education resource room, and whether the instructor is a special educator or technology educator will not markedly affect the design if the following principles are adhered to.

- Instructional units should be simple and not take more than one period to complete.
- Tasks should be varied and can alternate between hands-on and paper-and-pencil activities.
- Materials should allow the student to progress at his/her own pace (thus freeing the instructor to help others).
- Skills and products should be useful. This enhances the student's motivation and self-esteem. (Emphasis on such abstract manipulatives as assembly of nuts and bolts should be avoided.)
- Instructional materials should be portable, durable, and easy to store (breakdown and setup should be simple).
- Materials should not depend on specialized settings and require only one electrical outlet.
- Materials and tools should be related to those used in the regular program. Scrap material from the regular class can be consumed to minimize costs; used tools and equipment should not be overlooked, especially for teaching identification and safety rules.

Individual Self-Administered Instructional Units

Setting: This type of instructional material is best suited for a work-station setting. At each station the learner uses individualized instructional packages similar to those used in the occupational versatility program.

Method: The instructor acts as a facilitator, moving about the setting to assist individual students. This type of setting can easily accommodate seriously handicapped students with the addition of peer tutors or paraprofessionals who use the instructional material as a lesson plan. The disadvantage is the time (a minimum of one year) taken to make this sort of setting fully operational.

Example: An example of one form of instructional manual for one station is included in appendix A and pictures of the hands-on activity are included in figure 5–7. This instructional unit was

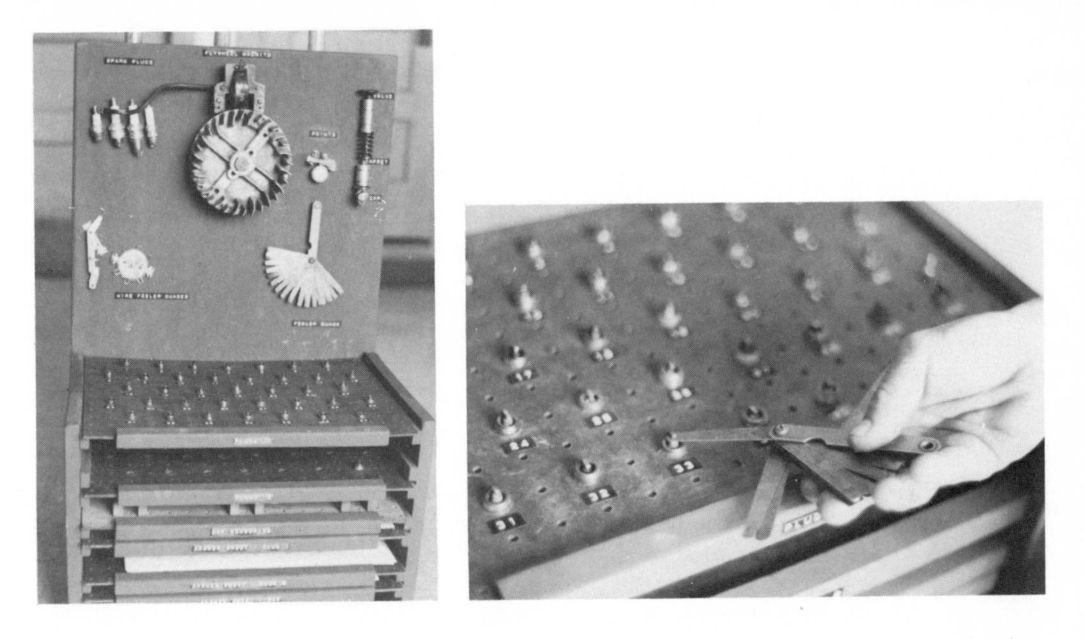

Figure 5–7. Individual self-administered instructional unit for fine measuring (related to small engines).

designed by a special education instructor at Olympia Washington High School. The sample is designed to teach the student fine measuring skills, which are needed in the small engine class. In this sample the student uses regular measuring tools (e.g., wire gauges, blades) to practice making critical measurements using real and simulated materials (e.g., spark plug gaps and gaps created by found hardware). The measuring skills are related to skills used in adjusting spark plug gap, point gap, valve lash gap, and flywheel magnito gap. The measurements require a precision of ± .001 inch in this unit. A complete work station related to mechanical power would include other manuals for engine parts assembly and disassembly, parts identification, tool usage and safety, use of troubleshooting charts or tests, and the use of technical literature.

Instructional Package: There are many ways to structure a learning package. One approach is summarized here in outline form. For another approach, the reader might consult the HERO Program in the home economics program at Tucson High School (see Babich 1980).

Instructional Package Outline

I. Directions for administration

   A. Materials and worksheets needed

B. Specific directions for maintenance

C. Directions for evaluating and instructing student (if needed)

II. Student Orientation

A. Package overview, brief introduction

B. Motivational statement

1. How the package relates to regular courses
2. How the package relates to jobs
3. How the package relates to current events of technology

C. Skills to be mastered

1. Technical skills
2. Work habits
3. Other

III. Administration

A. Directions to student
(Directions for students are outlined in a box while directions for the instructor are outside of box to make reading easier. See appendix B)

1. Pretest
2. Vocabulary needed
3. Work sheets needed
4. Any special directions
5. Specific directions for training

B. Student evaluation of interest in work performed

C. Post test

D. Optional advanced sections (research, problem-solving, etc.)

IV. Directions for scoring

A. Directions for scoring raw data

1. Norm tables, criteria for errors, directions to interpret results
2. Performance indicators

B. Directions for summarizing information (when information will be used in a vocational evaluation system)

Small Group Exploratory Instructional Modules

Setting: In this setting the teacher plays a more active role in instruction. The program can be based on the itinerant program established by Richard Michealis in Oregon. In this case, the instructor visits several high schools that feed the local vocational-technical school. Mr. Michealis visits the special education class and conducts prevocational instruction in the room for small groups of students. His particular program runs for thirty-six weeks and covers agriculture, metals, woods, electronics, mechanical power, and other areas outside of the technology curriculum.

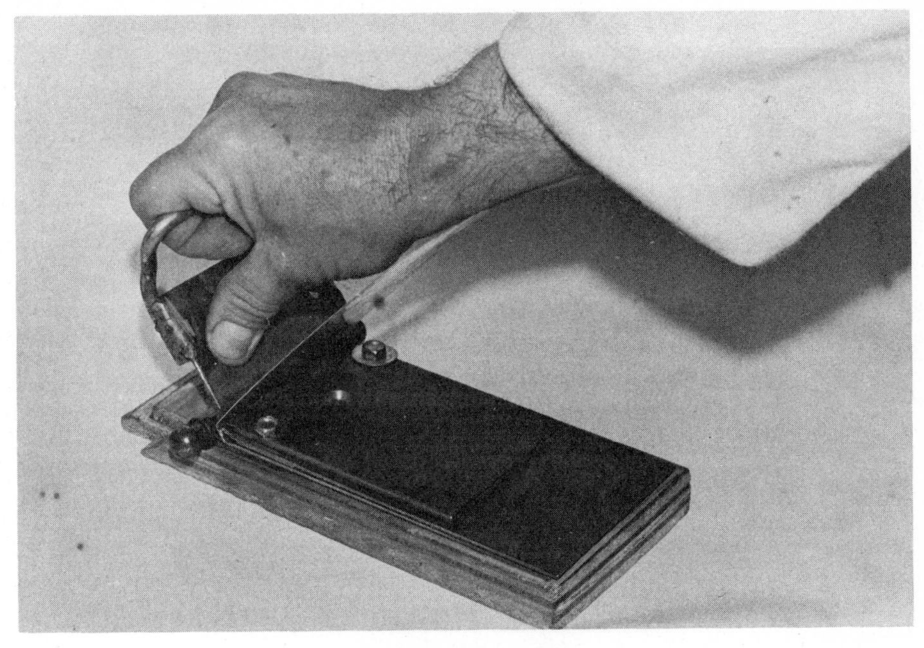

PORTABLE TOOL

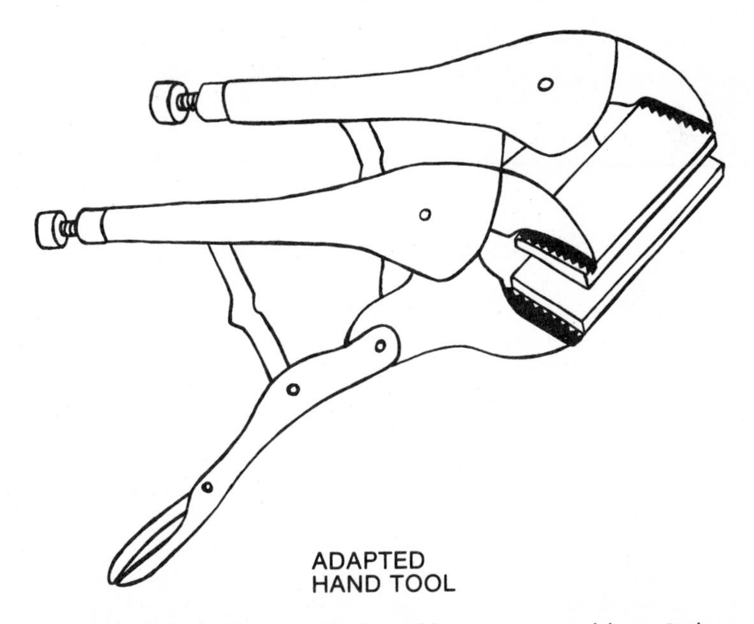

ADAPTED
HAND TOOL

Figure 5–8. *Two variations of a portable bar folder are presented here. Both are used in sheet metal projects. The bottom is constructed by welding two bars of metal to the jaws of vise grips. Both types embody the principles of design: low cost, use of scrap materials and old tools portability. (Illustration by Michael McHale and Todd Fly, Olympia (Washington) High School Drafting Students.Courtesy of Richard Michaelis.)*

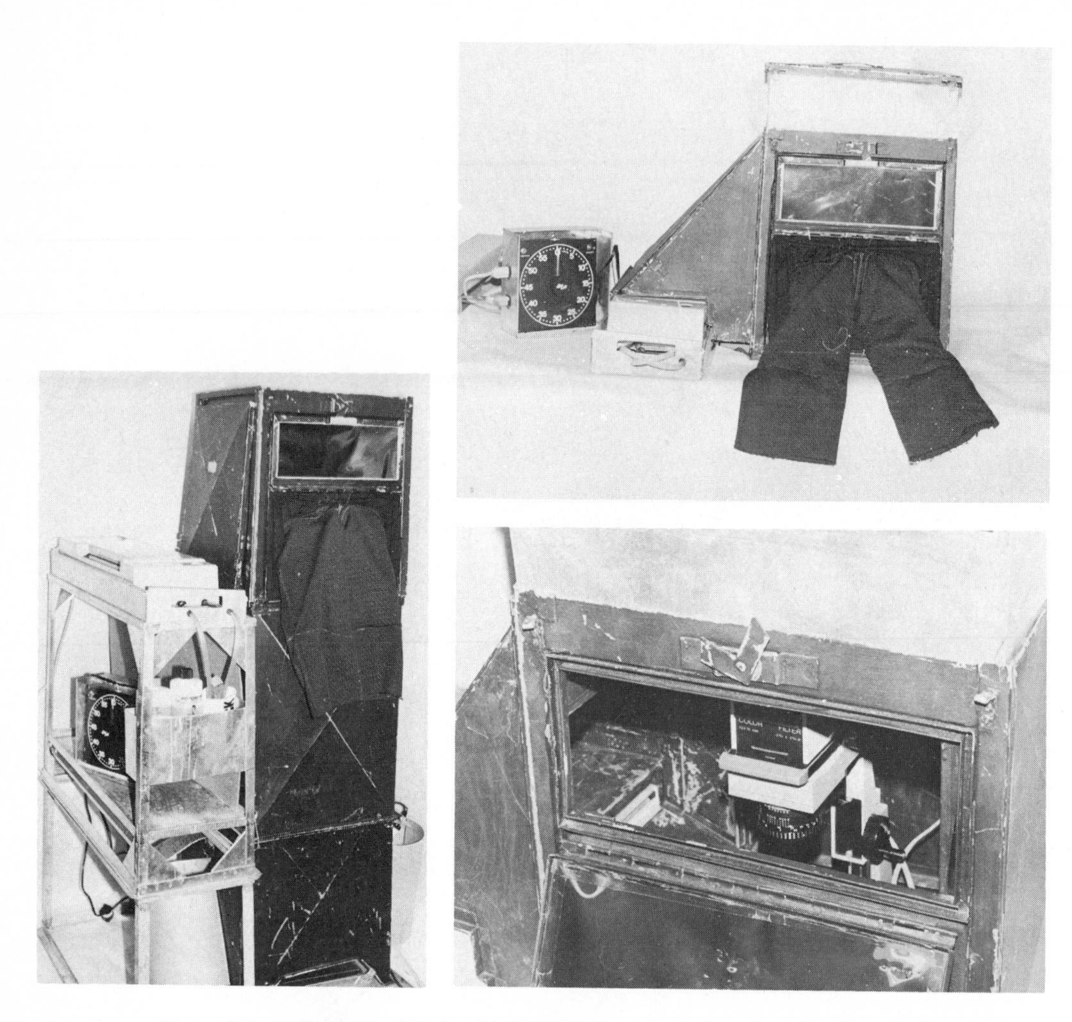

Figure 5–9. *(Courtesy of Richard Michealis.)*

A. Process camera and developing unit used in graphic arts mini-course can produce lithographic negatives and print pictures using an enlarger inside a light-tight box.

B. Unit is now set up for printing pictures. Student puts hands through sleeves to manipulate the enlarger and light-sensitive paper, which is shielded from light by this box.

C. Looking inside the black box, part of the enlarger is seen. This is what a student sees during orientation and practice.

Method: In small groups students are given demonstrations and background information to the technology being studied. Each module contains worksheets as well as a hands-on project or skill to be mastered. While one student receives direct supervision on the hands-on part the others are filling out worksheets. The disadvantage to this method is the small number of students

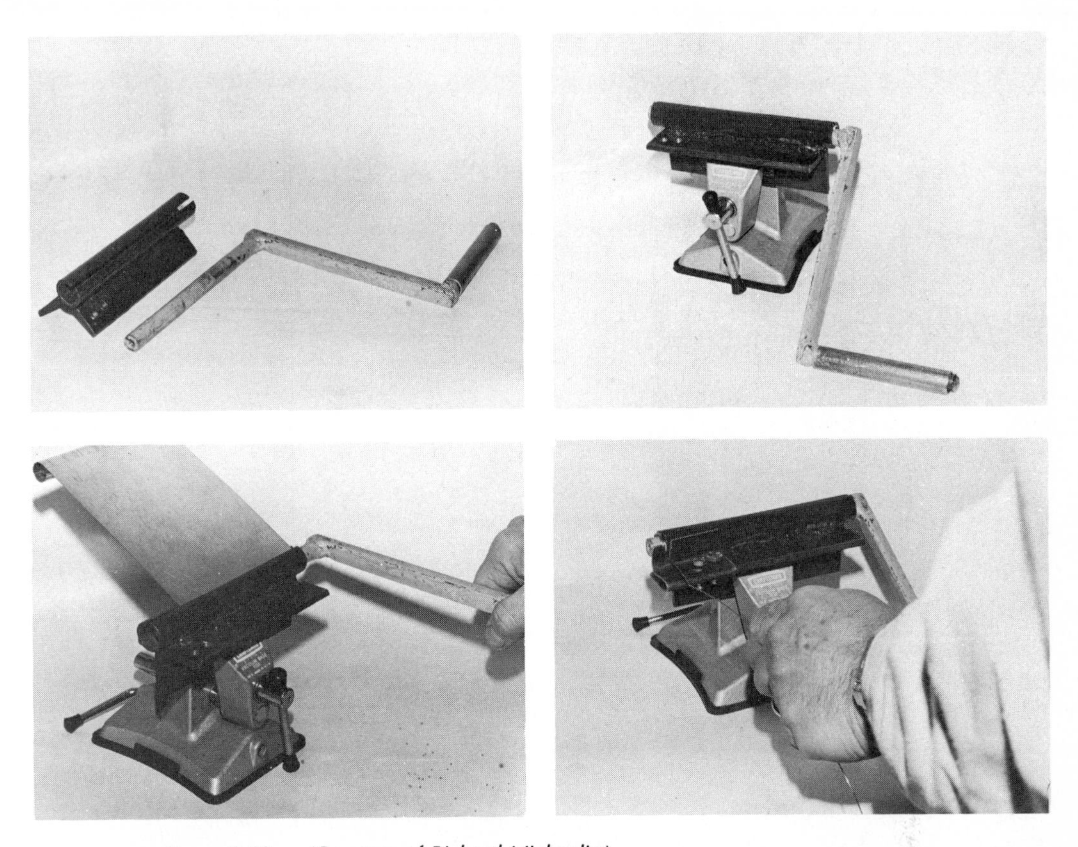

Figure 5–10. *(Courtesy of Richard Michealis.)*

A. These two pieces create the ferrule maker (see lesson plan in appendix B).

B. Assembled ferrule maker mounted in a vise that has a vacuum base for mounting on a regular tabletop.

C. Ferrule maker rolling an edge on a piece of scrap sheet metal.

D. Ferrule maker adapted for coiling a wire spring.

being served. However, the initial time spent in developing materials is less than in the individual self-administered instructional units. A combination of the two methods could solve the potential drawbacks.

Examples: Examples in this area are included in appendix B. Different aspects of this program are contained in the following figures and accompanying descriptions.

Worksheets to accompany the projects can also be designed with special needs learners in mind. A summary of the different types of worksheets would include:

- Applications/Writing skills
  Use of purchasing forms, bill of materials, job applications, questionnaires that progress from simple to complex.

Figure 5–11. *(Courtesy of Richard Michealis.)*

A. Small card holder assembled from two pieces with rolled edges. These were created with the ferrule maker.

B. Another project produced on the ferrule maker that can be used as a picture holder.

C. A collapsible picture holder.

D. A one-piece candy box. The rolled edges allow for assembly of the box without the use of rivets or soldering.

- Use of charts and tables/Reading and comprehension skills
  Use of catalogs and charts for feeds/speeds, fasteners, welding rods, troubleshooting, safety, etc.
- Interpreting design and use of codes/Reading and decoding skills
  Interpreting or generating blueprints, floor plans, operation sheets; the use of schematics, parts identification numbers, inventory codes, and computer codes.

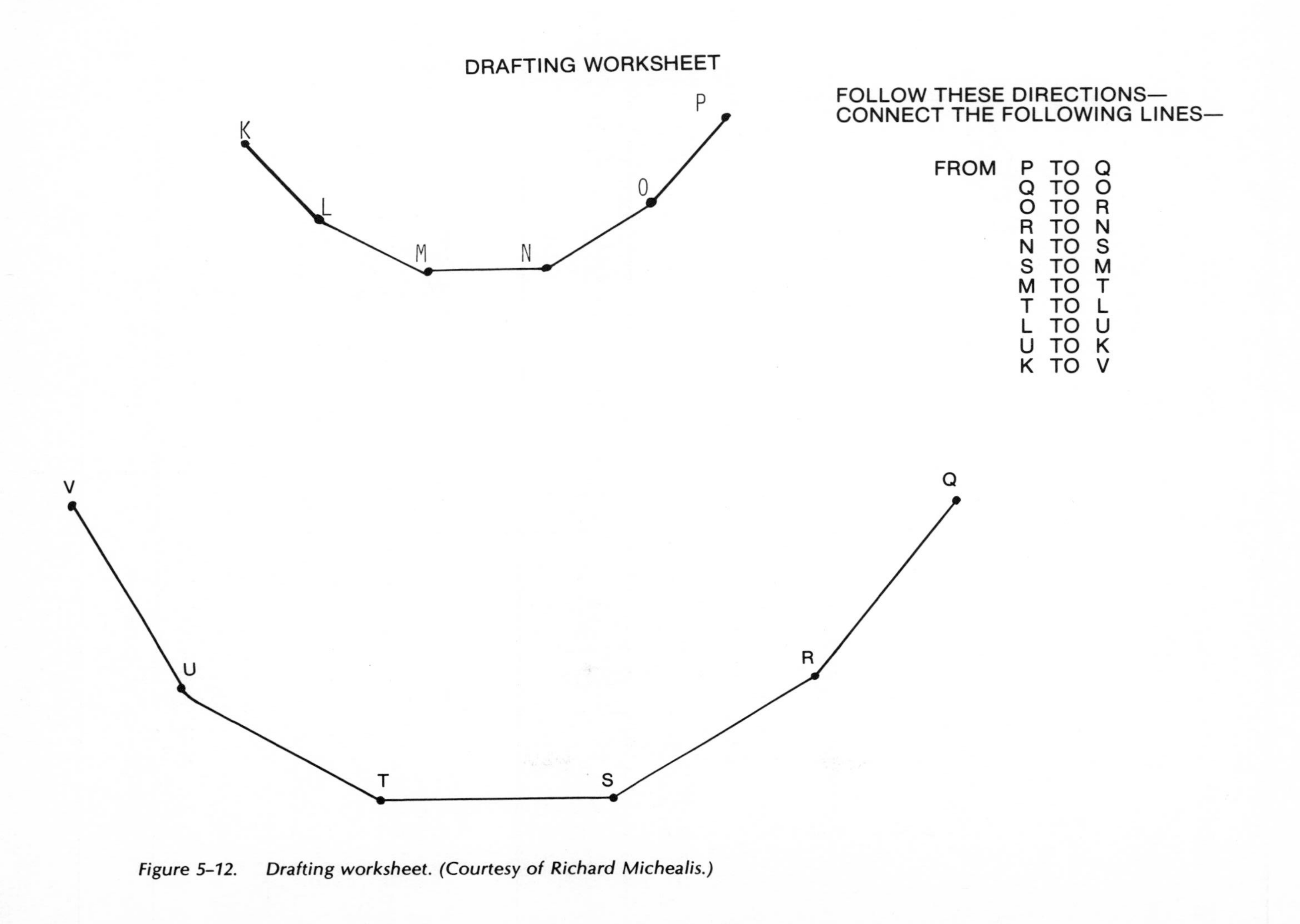
DRAFTING WORKSHEET
FOLLOW THESE DIRECTIONS—
CONNECT THE FOLLOWING LINES—
FROM P TO Q
Q TO O
O TO R
R TO N
N TO S
S TO M
M TO T
T TO L
L TO U
U TO K
K TO V
K
L
M
N
O
P
V
U
T
S
R
Q

Figure 5-12. Drafting worksheet. (Courtesy of Richard Michealis.)

LAY-OUT & DRILLING SEQUENCE:

Put the Numbers 1-6 in the circles in the order in which the step is done.
Fill in the blanks with the name of the step.

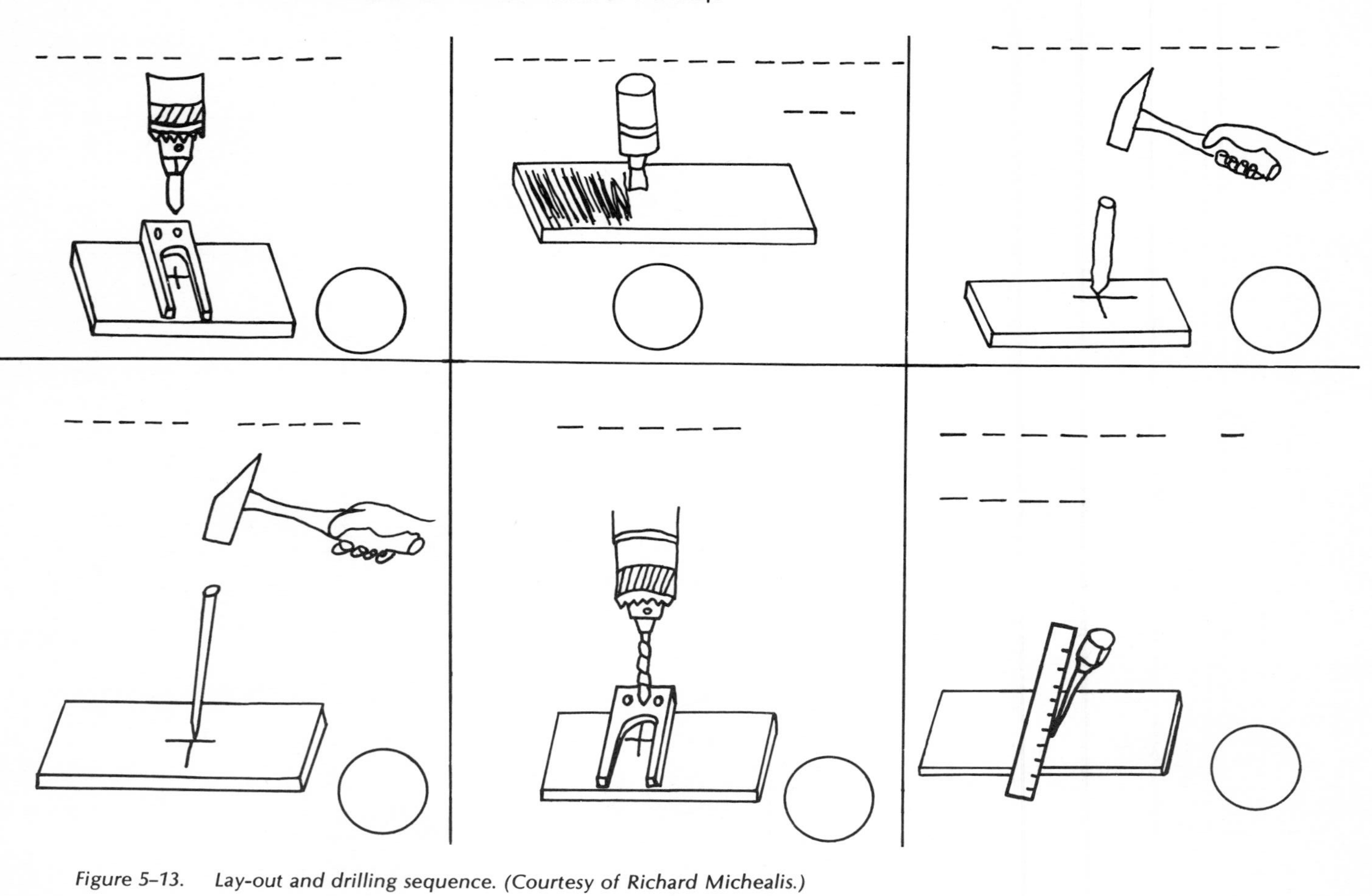

Figure 5–13. *Lay-out and drilling sequence. (Courtesy of Richard Michealis.)*

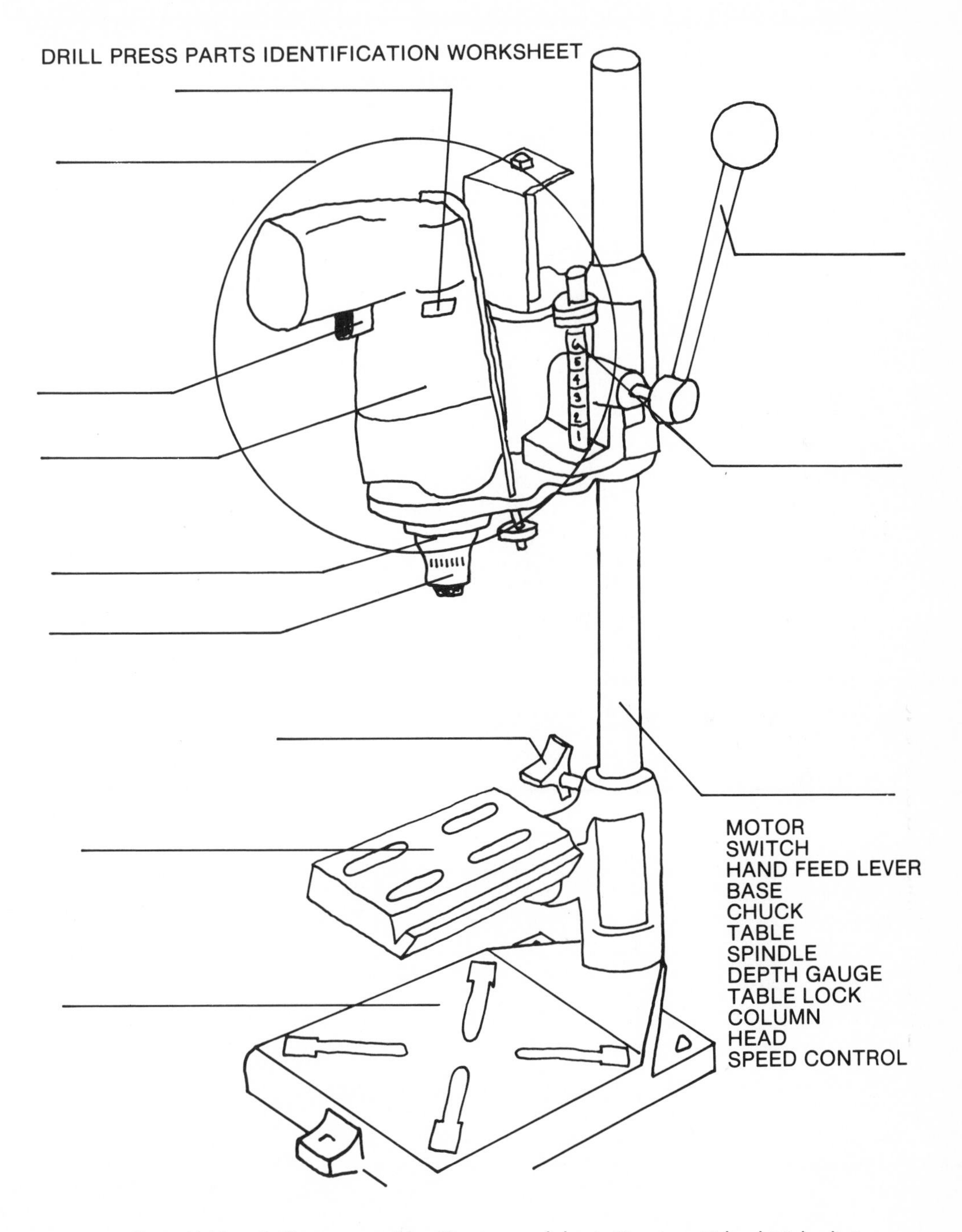

Figure 5–14. Drill press parts identification worksheet. (Courtesy Richard Michealis.)

1

WORK HABITS

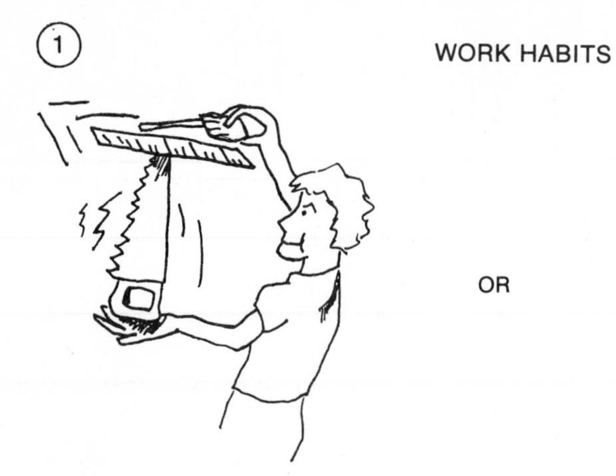

OR

2

OR

3

OR

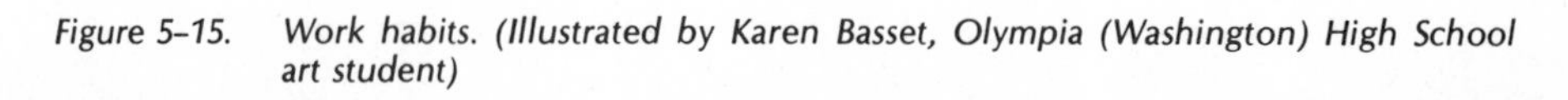

Figure 5–15. Work habits. (Illustrated by Karen Basset, Olympia (Washington) High School art student)

WORK HABITS

ANSWER SHEET

DIRECTIONS - 1) Match words to pictures.
2) Circle the best habit.

(1) ____ ____ ____ ____ ____ ____ ____ ____ ____

OR

____ ____ ____ ____ ____ ____ ____

(2) ____ ____ ____'____

____ ____ ____ ____

____ ____ ____ ____

____ ____ ____ ____

OR

____ ____ ____ ____ ____ ____ ____

____ ____ ____ ____ ____ ____

____ ____ ____ ____ ____ ____ ____ ____ ____ ____

(3) ____ ____ ____ ____ ____

OR

____ ____ ____ ____ ____ ____

____ ____ ____ ____ ____

____ ____ ____ ____ ____ ____ ____

- HORSEPLAY
- DIRTY
- WASH UP AFTER WORKING
- DON'T KNOW WHAT TO DO
- WORKING
- GET HELP AND ASK PERMISSION

Figure 5–16. Work habits, answer sheet. (Illustrated by Karen Basset)

WELDING SAFETY

Figure 5-17. Welding safety. (Illustrated by Karen Basset.)

- Following Directions
  Use of operation sheets, safety sheets, and worksheets related to instruction. Exercises in following directions, as in the drafting assignment found in figure 5–12. Learning how to sequence or follow a sequence, illustrated in figure 5–13.
- Identification of tools and machine parts
  This is similar to regular worksheets that involve identification except that it is simplified and follows hands-on application as soon as possible. See figure 5–14 for an example. Further simplification can be adopted from figure 5–13 where words can be matched to the number of dashed spaces representing letters. In this case a student with a limited vocabulary can complete the worksheet and begin to learn these words.
- Measurement/math skills
  Worksheets related to the creation of scales and their reading can be employed. Beginning with metric is often simpler than beginning with U.S. measurements. When using U.S. measurements students can begin with a drafting worksheet that starts with measurements at 1 inch accuracy and proceeds gradually to ⅛ inch.

WELDING SAFETY

ANSWER SHEET

DIRECTIONS: Match the number with the rule

NO GOGGLES ◯

NO GLOVES ◯

EYE PROTECTION ◯

GLOVES ◯

APRON ◯

BURNING ITEMS NEAR WELDING ◯

LEAKY HOSE ◯

SMOKE IN FACE, NO VENTILATION ◯

VENTILATION (FAN) ◯

COOL WELDS IN WATER ◯

KEEP GROUND CLAMP NEAR WELD ◯

Figure 5–18. Welding safety, answer sheet. (Illustrated by Karen Basset)

- Safety/work habits

  Worksheets can reinforce safety instructions as well as work habits. The adaptation for nonreaders (matching words to dashed lines) has been incorporated in figures 5–15, 5–16, 5–17, and 5–18.

Instructional Unit Design: This type of instructional format follows the same principles as the instructional package outline shown earlier. However, it is simpler in design as is evidenced by the following example.

Outline For Small Group Exploratory Modules

I. Activity: Single skill taught in isolation
   A. Tool use (e.g., ferrule maker)
   B. Parts identification
   C. Safety
   D. Uses of technology (film)
   E. Careers (film)

II. Worksheet: Applied academics
   A. Vocabulary game
   B. Measuring skills
   C. Design or plan sheet usage
   D. Test or worksheet related to activity in I

III. Activity: Skill leading toward terminal goal
   A. Layout of material
   B. Creation of sub-parts
   C. Practice with machines, materials, or tools

IV. Worksheet
   A. Worksheet that approximates worksheets found in regular classes
   B. Tests based on previous work

V. Activity
   A. Acquire a useful skill (e.g., gapping a spark plug)
   B. Produce a useful product (e.g., napkin holder, picture frame)

VI. Advanced part (optional)
   A. Visit a class related to module
   B. Participate in a field trip
   C. Research a topic
   D. Find book in library related to technology
   E. Instruct other students
   F. Modify product and create a variation

In appendix B are sample worksheets and possible lesson plan outlines.

COURSE OUTLINE FOR "BEGINNING WOODWORKING USE OF BASIC HAND TOOLS AND SHOP SAFETY" BY STEVE MILLER, DON SMITH ED. ALICE ROELOFS KREPS

The College For Living
Metropolitan State College

Introduction

A. Orientation
  1. Getting to know each other
B. Introduction to the shop
C. Safety Rules

Step I

Make bread board (Pretest)

A. Tools used
  1. backsaw
  2. mallet
  3. vise
B. Sanding
C. Finishing
  1. oil finish

Step II

Make mirrored wall shelf

A. Measuring
B. Tools used
  1. Metric rules
  2. Crosscut saw
  3. Rip saw
  4. Jack plane
  5. Files and rasps
  6. Screwdriver
  7. Coping saw
  8. Wood chisel
  9. Hand drill and bits
  10. Clamps
  11. Square
C. Sanding
D. Clamping
E. Gluing
F. Finishing
  1. Stain
  2. Varathane

Step III

A. Introduction to mass production techniques
B. Mass production project (see Figure 5-28)

Step IV

A. Student choice of a project

Figure 5–19.

Adapted Exploratory Class Projects

Setting: A regular technology class setting appropriate to exploratory-level class is used.

Method: This is the simplest type of curriculum to create. It need not be limited to the enrollment of special needs learners and as such can become the transition to mainstreaming or integration. Major accommodation consists of team teaching with a special education teacher, simplifying written work, and shorten-

MASS PRODUCTION PROJECT:

WHEELCHAIR TRAYS
DIMENSIONS IN MILLIMETERS

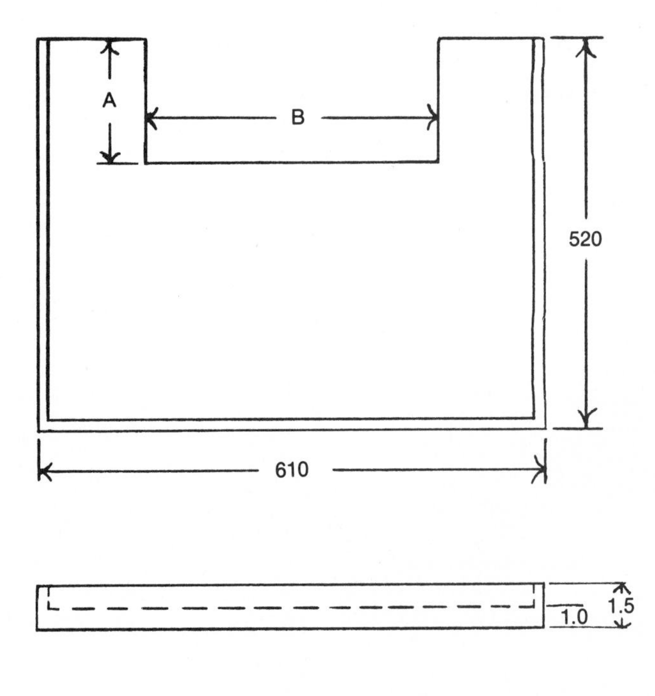

TINY TOT	A = 150	B = 285
JUNIOR AND ADULT	A = 200	B = 360

Figure 5–20.

Color Cues Help EMR Mainstreaming

By Michael A. Spewock

COLOR cues have proven to be a successful teaching strategy with educable mentally retarded (EMR) students mainstreamed in an industrial arts class. The model airplane shown here was the product of a mass production project involving eighth-grade boys and girls. Students with learning difficulties participated in the project by assembling the nine interlocking parts of the plane. [Editor's note: be sure your EMR students are not colorblind.]

Several weeks before the project was introduced to the rest of the class, the special needs students (under the supervision of the resource teacher) were practicing the proper method of assembly, using sample airplanes whose parts had been color-coded with felt tipped pens. A stripe of blue on the underside of the wing matched a stripe of blue on the top of the fuselage. A stripe of red on the tail section matched a stripe of red on the rear of the fuselage. A stripe of green on the propellor matched a stripe of green on the nose of the plane.

When students demonstrated competence in assembling the color-coded components, the cues were faded; that is, instead of a full stripe of color, a small dot of color was used on matching parts. Students then practiced assembling and disassembling the airplane until they again showed confidence in their performances. Finally, students were allowed to practice assembling plane parts that provided no color cues at all.

When the industrial arts class began production of the model airplane, mainstreamed EMR students were ready and able to join their classmates as workers on the assembly line. □

Michael A. Spewock is a graduate assistant, Industrial Arts Department, The Pennsylvania State University College of Education, University Park.

This model airplane, with nine interlocking parts, was a mass production project for eighth-graders. EMR students were able to assemble it after practicing with color-cued models.

Matching parts are color coded. Blue stripe under wing matches blue stripe on top of fuselage.

Green stripe on propellor matches green stripe on nose of airplane.

Red stripe in tail section matches red stripe in rear of fuselage.

Fig. 1—Use of Color Cues

Figure 5-21.

ing instructional units. This form of instruction lends itself easily to small group and/or large group instruction. Peer tutoring becomes a complementary teaching and learning experience. Special and innovative ideas should not be overlooked. For example, home repair, industrial arts/crafts, interdisciplinary clusters, work co-ops (e.g., make jigs and fixtures in wood class for mass producing root beer; the proceeds can be used in units on community and daily living).

Micrometer Reading Kit

by Samuel L. Skeen

An easy way to individualize the teaching of micrometer reading

Each year hundreds of instructors in our schools teach thousands of students to read the micrometer.

Among these thousands are some who do not grasp the concept the first time it is taught. If you have had students in this situation and desire a simple yet economical method of giving individualized instruction and practice, the teaching aid pictured here will meet your need.

Operating on the principle that the facing card cannot be removed unless the rod in the photograph is placed in the slot corresponding to the correct answer, the unit gives the student immediate feedback. When the correct answer is selected, the card is pulled from the box and placed back in the rear of the stack of cards that have already been made for use in this kit or it is laid aside. The student then repeats this process until all cards have been read.

Before the holes are drilled in the box, the 5″ x 7″ cards should be prepared. A drawing of a micrometer sleeve and thimble (as shown in the photograph) is made on each card. The various graduation marks on the sleeve and thimble are labeled as you desire, and four readings (only one of which will be correct, of course) are listed in multiple choice fashion. Then all cards are placed in the box and the holes are drilled. After that, cut a slot in the card corresponding to the correct answer, and the unit is then ready for use by the students.

Copies of the drawings can also be made, labeled, and used as masters for student worksheets. ■

Mr. Skeen, an instructional materials coordinator of an area vocational-technical center, lives in Ripley, W.Va.

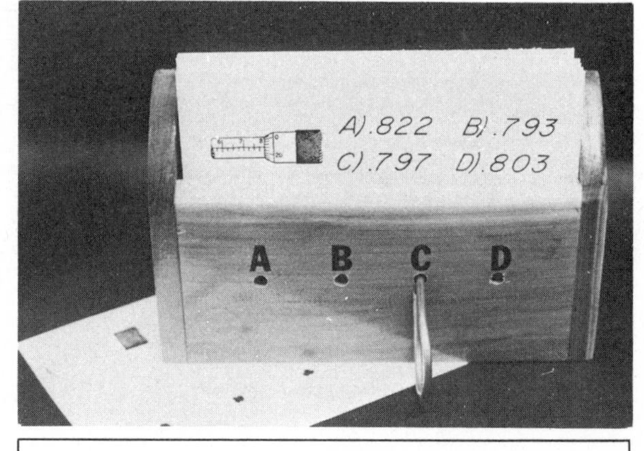

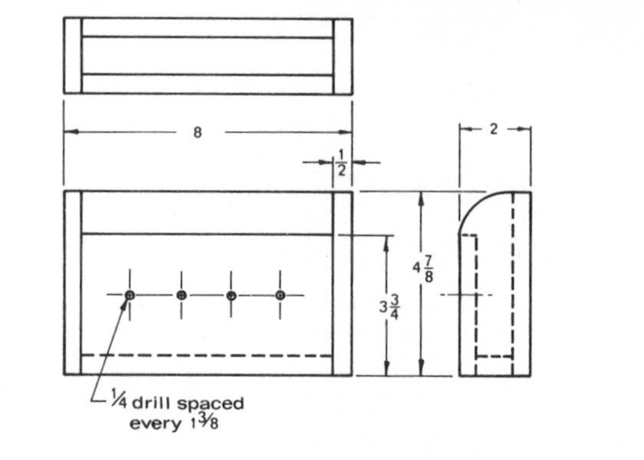

Figure 5–22.

Two-Cycle Engines In Peanut Cans

by Daniel L. Stone

How one teacher gets his students started on their work with small engines

Have you been using four-cycle small engines for hands-on activity in your power program (and then relying on a couple of tranparencies and handouts to cover two-cycle theory)?

Does your beginning drafting or design class need a sketching exercise with which to sharpen their skills and knowledge of working and assembly drawings?

Are you tired of textbook drafting problems? Are you tired of searching for an object that would require multiple plates to describe in an advanced drafting class?

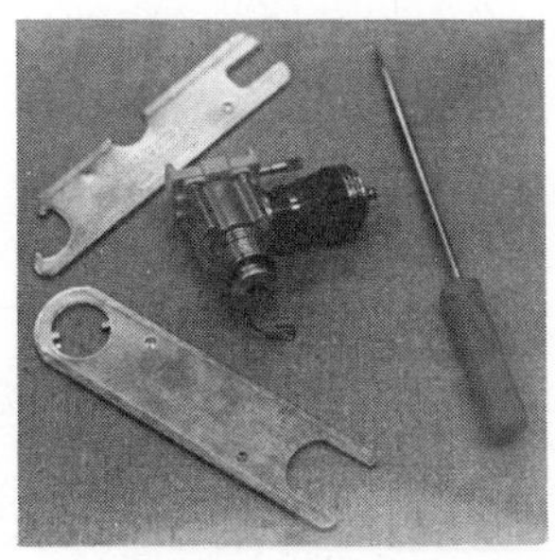

Engine and tools needed to work on it.

Everything needed for a class of 20 students fits into a medium-sized box.

Read on if you answered yes to any of these questions. Because by shifting the emphasis on the information in this article, you can help yourself (and your students) with any of these objectives.

I have been using the Cox model airplane engine for two years now with my 7th-grade power classes. Their work with it takes about 10 days and is an inexpensive and quick way to get the students actively involved with two-cycle theory.

The engines, which most students are familiar with, are small and have a minimum number of parts. There are only three tools needed to work on them and everything fits very nicely into a small peanut can.

By teaming two students per peanut can "kit" you can fit everything you will need for a class of 20 into a medium sized box. And this can be important if your storage space is at a premium as ours is!

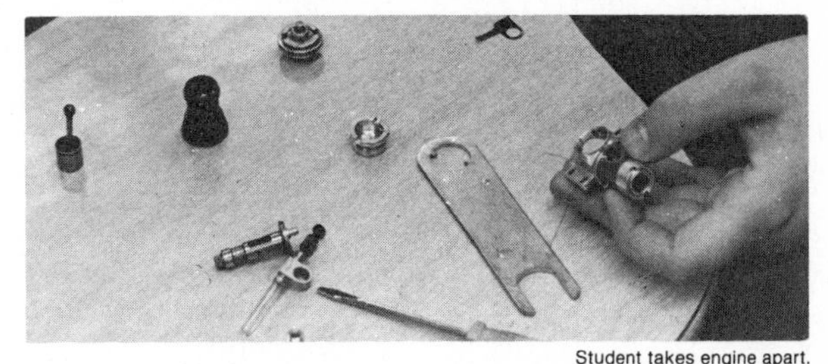

Student takes engine apart.

What we do

Every student is given a blank composition book. The students are grouped into teams of two, and each team is given an engine "kit." During the next two weeks we turn the composition book into a two-cycle engine booklet. The booklet has three sections. Section one is the theory of operation and consists of notes from lessons and a section of Cox's booklet "How To Get The Best Performance From Your Model Engine" on engine operation. (This booklet is available in classroom quantities from L. M. Cox

Figure 5–23.

Day-To-Day Activity

Day 1 —Introduction to activity.
—Divide class into teams.
—Composition books and kits issued and examined.
—Class discussion of uses for two-cycle engines.

Day 2 —Engines torn down and parts identified.
—Two-cycle theory: cross flow and loop scavenged.

Day 3 —Cox two-cycle booklet read.
—Complete section one of composition book on theory of operation.

Day 4,5,6—Complete section two of composition book, working drawings of actual parts, and descriptions of functions.

Day 7,8 —Section three completed, exploded assembly drawing.
—Booklets in for grading.

Day 9 —Two-cycle engine, care and feeding.
—Troubleshooting.
—Engine reassembly.

Day 10 —Run trial.
—Troubleshoot as needed.
—Prepare engines for storage.
—Collect engine kits.
—Return composition books.

Mfg. Co., Inc., Service Dept., 1505 E. Warner Ave., Santa Ana, CA 91705.) Section two is a working sketch of each part and a written description of the function of each part in the engine. The descriptions come from lessons and the parts drawings are made from the actual engine parts in the students' kits. Section three is an exploded assembly drawing to show how all the parts fit together.

After completing the booklet, the students take their engine apart, reassemble it, and try running it. If it runs, it is oiled, returned to its can, and turned in. If it does not run, the student has to troubleshoot it until it does run or until he can explain why it does not. (Most often the problem is with a reed valve.)

We grade the activity by averaging their booklet grade and the grade for their laboratory work on the engine. In the booklet I look for neatness and ability of the student to follow directions. Laboratory work places emphasis on the correct use of tools, proper treatment of the engines and parts, and how the students work with each other.

Other advantages

Besides getting students to work right from the start, this activity has other advantages. First, it helps to settle the students into your routine. Secondly, it gives you a chance to gauge the students' abilities before you turn them loose on the four-cycle engines and all the different tools they will need. And thirdly, after reading the Cox booklets and working on the engines, the students start developing an appreciation for this segment of their technological world.

If your class is drafting or design rather than power, you need only shift your emphasis from the theory and running of the engines to the drawings themselves, and you will have a motivating activity.

For a design sketching activity, you could stick with the composition book. You would want to play down the theory and emphasize the sketches. Depending on the group's ability, you might want to introduce shading at this stage also.

A drafting class would be making separate plates of the various engine parts. Depending on how involved you want to get, you could develop a handout with parts' sizes for dimensioning. Or going to the other extreme, get them involved with precision measurement. Borrow or requisition a few calipers, rules, micrometers, and hole gauges, and the students can start developing a hands-on appreciation for tolerances, fits, and the concept of interchangeability of parts. ■

Mr. Stone is Chairman of the Industrial Arts Department in Abington Heights Middle School, Clarks Summit, Pa.

Figure 5–24.

Examples: Examples for projects shown here were selected from existing materials found in journals or classes involved with the study of technology.
Instructional Methods: See the next section on regular class placement.

REGULAR CLASS PLACEMENTS

A major theme in this book is that mainstreaming and least restrictive options can only be realized when the learner can progress from the most restrictive and most specialized to the least restrictive and most normalized setting within a coherently structured educational framework. In other words, the student or individual teacher cannot hope to maximize educational opportunities alone. It involves the special educators, who help prepare the students for placement in regular classes and provide support services for the students and teachers. Among the services they can provide are:

- Coordination (or case management) of student instructional services and placements (e.g., vocational evaluation, IEP, cooperative instructional agreements)
- Instructional support (e.g., tutorial service, supplemental instruction, adaptation of instructional material and technique)
- Planned integration process
- Links with postsecondary programs

Without these types of components, placement in regular classes becomes potentially haphazard and the benefits of normalization cannot be fully realized.

The implementation of these services will hinge on changed attitudes toward handicapped persons and toward one's profession. Special education teachers must be wary of their overuse of specialized jargon and forms that may make the blending of two fields forbidding. They must recognize the legitimate frustrations of regular teachers and assist in all ways possible. Regular education teachers must re-examine their attitudes toward students and their commitment to all learners in light of their existing instructional methods and materials. And administrators must pay more than lip service by providing the necessary time for planning, material development, and professional growth.

Cooperative Instructional Agreements

The picture is coming into focus. The vocational educator is skilled in a craft and knows what the job market demands of entry level workers in

> that occupational area [are]. The special educator is skilled in working with handicapped students, especially in regard to acquiring basic skills and an improved self-concept. The two complement one another perfectly if they will only talk and particularly listen (Boland, 1979, p.10).

The concept of a cooperative instructional agreement between the special educator and the technology instructor is not a new one. The specific term was used by Alan Phelps (1975) in an in-service resource guide. In this guide he stated the need to "systematically coordinate personnel, resources and services. . . [where] the greater the severity [of the handicap], the greater the need for coordination of multiple services." Penny Holeman, a work experience coordinator in the Portland (Oregon) Public Schools, simplified this concept to a process of establishing rapport between instructors. She gave as examples the need to establish a working relationship so that teachers could evolve, in advance, a mutually agreed upon procedure for responding to the needs of students.

The entire process can be elaborated upon and formalized into a mini-in-service in which this "rapport" can become formalized as a standard operating procedure. This process would extend the interdisciplinary team effort established during the IEP conference (see chapter 2 for IEP descriptions). As the student finds his or her way within the regular class (and the instructor feels more confident), this formalized process can be faded out. This cooperative venture can be developed in four major steps.

Step 1: Establish Rapport and Familiarity

The purpose of this step is to become aware of each other's teaching environment, techniques, and students. Some practical ways of establishing these are:

1. Secure release time to visit the alternate class and serve as an aid.
2. Participate jointly in organizational meetings and conferences.
3. Conduct mini-in-services after school about specific topics that describe one's program (e.g., IEPs, safety, discipline, and grading procedures).
4. Review past efforts at mainstreaming or cooperation, listing strengths and weaknesses.
5. Share articles from journals, team teach, try joint afterschool projects (e.g., Special Olympics, Vocational Industrial Clubs of America VICA, visit training centers)

The results of these activities can produce the type of enthusiasm and increased understanding expressed by a teacher in the Madison (Wisconsin) Metropolitan School District:

> I find the regular teachers have come to respect and admire me for what I can do. They understand why a special education class must have a richer

teacher-to-student ratio. . . . On the other hand, I am impressed with the wide grasp of knowledge that a regular teacher can bring to a subject. I'm learning as much from them as my students are. (President's Committee On Mental Retardation, MR 78)

Step 2: Determine What Types of Information Will Assist The Regular Teacher

As in the IEP meeting (or in lieu of it), certain types of information will be useful to share with the regular teacher regarding the special instructional needs of the student. Two generally broad types of information will be useful: a profile of the student's abilities and needs (rather than labels), and a rationale regarding the placement in the regular class.
Examples of the type of information contained in these two categories would be:

- Profile
  - Academic abilities (reading, writing, math, and study skills)
  - Training/instructional methods the student benefits most from
  - Degree of motor coordination (if exceptional and related to course requirements)
  - Personal adjustment needs (how the student gets along with peers/teachers, ability to handle stress/frustration, maturity level, special considerations)
  - Medical/Health considerations
- Rationale:
  - Interests of parents, student, and teachers in class
  - Goals of placement: skill acquisition, personal adjustment, regular class placement
  - Goals of modified placements with special objectives or conditions
  - Goals for regular placement

Step 3: Communicate about Specific Special Learning Needs As Warranted by Placement

In addition to the information mentioned in step 2, it may be important to supply detailed descriptions of the student's specific needs. Because chapter 3 gives background on disabilities, these needs are stated here in educationally relevant terms. Examples of these can be:

- Special learning needs: specific instructional needs such as textbooks with simplified vocabulary or alternatives to reading, assistance needed in measuring, extra time needed for memorization.
- Special hearing needs: placement in classroom, monitoring of hearing aid, use of tutors and notetakers.

- Special vision needs: need for low-vision aids, accessibility and mobility training, audiocalculator, magnifying glass, lighting sensitivity.
- Special orthopedic needs: accessibility and study modifications, special equipment or tools, who to contact for therapy consultation.
- Special communication needs: alternatives to spoken language, what to do when teacher and student cannot understand one another, specialist to consult.
- Special behavior needs: signs that indicate a potential behavior problem, what is reinforcing, behavior programs/contracts the student participates in, medications and effects on coordination or learning, how to respond to specific forms of behavior.

Step 4: Delegate Responsibilities

Once steps 1, 2, and 3 are in place, what will happen during the placement needs to be worked out. A game plan is given here that should be constructed on an informal and flexible basis. Questions that might be useful in determining the specific instructional responsibilities are:

1. Who will adapt existing writing and reading assignments?
2. How will information from different professionals working with the student be shared? (This decision involves the student and parents as well.)
3. How can parents be kept informed? Who will take the responsibility?
4. Who will keep administrators informed?
5. What should be the contingencies if the student fails to maintain acceptable behaviors? (See chapters 3 and 4.)
6. If the student cannot benefit from existing instructional methods and materials, how can instructional efforts be pooled?
7. Is there a need for periodic staffings or briefing regarding the student?

If the student comes into the school new, if there is no vocational evaluation report, and if the IEP lacks an interdisciplinary composition, then many of the items in steps 1 through 4 will be difficult to answer. One way of gathering the information is to provide an entry-level course as a starting point. Another way involves the deployment of a pre/post test that represents the technology curriculum. These can be built around simplified versions of entry-level instructional material or the informal vocational evaluation described in chapter 2.

The Madison (Wisconsin) Metropolitan School District has formulated the following guides and teacher objectives for the Basic Class, which is team-taught as an entry-level course. These types of objectives reflect cooperative instructional efforts and delineate the different instructional responsibilities. These are summarized as:

1. The team reviews the courses to identify the success of the curriculum in mainstreaming students into regular vocational classes.
2. The team identifies student needs in a vocational area.
3. The team identifies and implements various teaching techniques used to meet exceptional educational needs and their effectiveness is reviewed weekly.
4. The team implements evaluation of learner activities daily (or when appropriate).
5. The team makes recommendations for future courses and discusses them with the student and the counselors.
6. The team shares information about the curriculum with parents and keeps them informed at least once each grading period.
7. The team has monthly staff meetings to evaluate the program and share ideas.

An individual student instructional plan written for a technology cluster could include the information found in figure 5–25.

Tutors

A tutoring system is often established by using peers as tutors. The benefits of such a system are well documented by teachers, tutors, and the students receiving the services. A good tutoring system will not blossom overnight. It is the product of planned recruitment and training and requires a concerted effort to maintain it. The dividends from such a program are large in comparison to the initial investment of time.

Benefits of Peer Tutors

The special education student who accepts and utilizes peer tutors will obviously benefit from an instructional standpoint. In addition, this is bound to carry over into the student's self-concept since a positive self-image is built upon successful performance in class. Besides the tangible benefits of one-to-one instruction, the exceptional student is provided with a chance to socialize with a peer student, which may later enhance acceptance by other students. Some studies suggest that learning disabled and mildly retarded students learn more efficiently with peer tutoring than with small group instruction. Many students may eventually progress from student to peer tutor themselves.

The benefits of tutoring another student are likewise well established. Having to teach another student moves the tutor toward a more complete

CLUSTER/PROGRAM: WORK TRAINING

Id: No.: 1

Task: Clean and adjust spark plugs

Progress: Introduced	Progress: Involved	Progress: Productive	Progress: Employable	Occupational Performance Objectives	Basic Skills/ Concepts	Basic Skill/Concept Content	Pro-gress: Introduced	Pro-gress: Developing	Pro-gress: Competent
				Given the necessary tools, materials, equipment, and requisite knowledge, the learner will:					
				1. Pick up and place plug in solvent solution	Finger dexterity	Grasp and hold spark plug			
				2. Remove and place plug in plug holder jig with bottom up	Form discrimination	Recognize electrode (top) and ground (bottom) ends of spark plug			
				3. Wire brush until clean (use cleaned condition picture for comparison)	Hand-eye coordination	Brushing strokes			
				4. Determine plug gap by using different sized gauges and placing them between the gap until appropriate size is located	Number recognition	.025, .030, .035, .040 (sizes on the various gauges)			
				5. Determine whether plug gap is correct by comparing gauge size with the master size on the poster attached to the bench	Number recognition and matching	Match 3 digit numbers shown above with those on a wall chart			
				6. If necessary, close gap by tapping the ground of the plug lightly with a soft-face hammer until the gap gauge is moveable within the gap, but both the electrode and ground are touching the surface of the gap gauge	Grasping Fine motor coordination Manual dexterity Touch discrimination	Hammer handle Light tapping of hammer on metal Feeling the distances between metal surfaces by placing a metal plug between the surfaces			
				7. Remove plug from jig and place in inspection basket	Grasp and remove				

Occupational Instruction	Basic Skill/Concept Instruction
Teacher Activities: Design and build a spark plug holding jig that will hold plug in stationary, inverted position Provide individual demonstrations of each step in procedure Instructor places hands on learner's hands and manipulates fingers for various steps Drill the learner on recognition of single numbers, then multiple numbers Continuous reinforcement	***Teacher Activities:***
Learner Activities:	***Learner Activities:***

INSTRUCTIONAL RESOURCE MATERIALS

Name/Title	Media	Source	Name/Title	Media	Source
Spark plug holding jig Spark plug gap gauge Number charts Wire brush Spark plug cleaning solvent Soft-face hammer					

Figure 5–25. Cluster/program: work training (Courtesy of Alan Phelps from material originally appearing in Instructional Development for Special Needs Learners: An In-Service Resource Guide *published by the Illinois Network of Exemplary Occupational Education Programs for Handicapped and Disadvantaged Students, 1975.)*

mastery of the material. Evidence exists that supports the idea that low-achieving students learn better when they are placed in the role of teacher than when they study alone. Non-handicapped tutors additionally benefit by broadening their acceptance and empathy for similarities and differences all persons share. Finally, there is the building of esteem and status when a student becomes a tutor. This is especially true for students who have a low self-concept to begin with. It becomes an opportunity to make a contribution, to demonstrate abilities.

An example illustrating this comes from the first case study described in this chapter. In this case study several students in an electronics class volunteered to tutor students from a special education class. Many of the volunteer tutors had never previously distinguished themselves in regular class performance, and, in fact, were receiving tutorial assistance themselves under a Title I program. However, during the two-week mini-course involving peer instructors and handicapped students, the attendance of the peer tutors improved and they demonstrated abilities and skills that had previously gone unnoticed. Thus, the tutorial program offered a valuable alternative to the regular course instruction and assessment methods.

In this example the peer tutors provided one-to-one instruction that could not otherwise have been provided in the electronics class. The tutors also served as role models for appropriate personal behavior and safety practices. Additionally, the peer tutors seemed to be able to motivate and instruct where regular teachers had experienced some difficulty. The impact this type of experience had on the tutors is documented in the following answers taken from their final examination:

Question — How has your attitude changed toward people with handicaps?

Answer #1 — Working with handicapped students has been interesting and enjoyable. When we were first introduced to this program we didn't know what it was all about. We weren't sure if we wanted to spend the time . . . . after a few visits we got to know the other students and they got to know us . . . . we really felt like we were helping out.

Answer #2 — I thought they would be abnormal. I thought they wouldn't be able to do things that are required in electronics. But you teach them, and show them and keep on showing . . . they will catch on and do it themselves . . . . my attitude has changed . . . . In my opinion I think all of them can learn to do something, they just need time and a teacher.

Answer #3 — I like teaching because I learned to accept mentally handicapped people more than I did before . . . sometimes it was hard, sometimes easy . . . it's the mood you or your students are in . . . it will give me a little more respect for the teacher.

Answer #4 — I thought they would be stupid . . . couldn't learn anything. I was not sure at first. . . . I found that once I got to know them they were easy to get along with. . . . I like it because it gave me a better understanding of handicapped people and helped me to realize how fortunate I am to be such an "intellectual whiz" . . . . teaching certainly is difficult, a lot of patience is required. . . . I'm surprised that I had it.

Sources for Tutors

Where can you recruit tutors? Before answering this question it is useful to distinguish between a formalized tutorial program (often found in special education programs) and an informal one (usually associated with regular classroom settings). In a formal program the tutors are often recruited from the student body at large in a standardized manner. In the author's case this included interviews, applications (figure 5–26), and recommendations from counselors. During the interview the student receives an orientation to the program requirements, practices, and grading procedures.

An informal tutorial program often recruits from within the class. This has several advantages because it usually means tutors are already familiar

STUDENT AID APPLICATION
FOR VOCATIONAL EVALUATION

NAME____________________________ PHONE______________

GRADE__________ PERIODS FREE________________________

1. Do you know what we do in vocational evaluation?
2. Why do you want to volunteer here?
3. Answer the following professional questions:
   a) Have you ever worked with students in special education?
   b) What are your attitudes toward students in special education?
   c) Can you be confidential in your work with these students?

      ______ Yes   ______ No

   d) Do you have patience when teaching?
   e) Do you agree to show up every day?
      To notify us when you can't?
   f) Are you willing to correct papers and work with tools?

CRITERIA FOR CREDIT: 1. Training or intake
2. Short one-page report or work sample
3. Be present 90 percent of the time

I agree to the above criteria for my grade.

____
Student Signature

Figure 5–26. Student aid application for vocational evaluation

with the on-going program. They can be recruited for a variety of reasons, as listed below:

1. The tutors may be students who need an alternative to regular assignments for purposes of make-up work or supplemental work.
2. Students late in entering a course and who are unable to keep up or function with the regular class.
3. Advanced students who can benefit from an enrichment activity. Often these students need to explore their own prejudices toward people with different or lesser abilities. Another aspect is the exploration of service occupations related to technology.
4. Use of existing cooperative assignments. Most classes employ a buddy system. In Denver, Colorado, Doug Woolverton, an auto instructor, uses teams or crews of three-students in which the advanced students serve as crew chiefs responsible for guiding and instructing the other members (of whom one is often a special education student) (Boland 1979). In the Basic Class (previously described) one of the student objectives is to complete his/her share of the work while participating on a two or more member team and to ask his/her partner questions before asking the teacher.

When recruiting for either a formal or informal program, the initial selection of tutors deserves some thoughtful preparation. Experience shows that some students may opt for a tutorial assignment to avoid completing regular assignments. At other times tutors sign up out of boredom or lack of options in the regular curriculum. Students with these motives can be screened out most of the time if the recruitment accurately portrays what is expected of tutors and if the tutors must furnish some evidence of sincerity or motivation (e.g., attendance, counselor's reference).

Training Tutors

In the informal approach, training in less intense. Generally, students receive the following types of training:

- Introduction to handicapping conditions, stigma, the rights of handicapped citizens, the important and sensitive role of the tutor
- Behaviors expected of tutors (e.g., friendship, instruction, reporting, suggestions)
- A structured approach for instruction that the tutors must master through observation and teacher supervision (e.g., task analysis, data collection)

For more formal programs the above items are often presented in greater detail along with field trips, reading, films, and speakers. In addition, training may touch upon:

- Practice in specific instructional techniques (see chapter 4)
- Design of instructional materials
- Discussions about normalization and peer role models
- Paraprofessional/professional careers and responsibilities
- Practice with specific skills needed in the technology lab

In some instances the initial recruitment and training can be centralized in either the special education or technology education department. Some programs train their aids after school, at lunch, or in special classes and clubs. Do not neglect consideration of outside experiences that can enhance and broaden the experience. Examples of these might include reading books, participation in outside recreational activities by tutor and special student, and the creation of special projects such as films and displays.

Maintenance of the Tutorial Program

Tutorial programs require constant maintenance. This maintenance revolves around tutor accountability and recognition of tutorial efforts. Accountability can be maintained by a simple log of activities or the use of a more structured checklist, as illustrated in figure 5–1. This can be combined with brief supervision or observation of the student tutor's performance to note such aspects as:

Attendance and punctuality
Recording of necessary data/reporting
Acceptance of criticism and suggestions for improvement
Follow-up on problems in instruction/initiative
Whether or not student being tutored likes peer instructor

In addition, some type of experience should be provided for the student to share his or her experience and expertise. These could include:

Conducting an open house
Talking before other classes and recruiting other tutors
Presenting a lesson and giving tests
Debriefing sessions, seminars, or group discussions

The efforts of the tutors should never be taken for granted. Some suggestions for recognizing the tutors are:

Invite the school newspaper to do an article on the tutoring program

Call or write letters to parents and counselors regarding performance

Award certificates or letters of reference

Have the students receiving the service evaluate the tutors or plan a social event to show their appreciation

Ask the students to summarize their experiences and to critique tutorial program.

Integration of Peer Tutors

Some students might view the peer tutor's help as a source of stigma. Likewise, tutors should not be imposed on students. If the tutorial program is talked about from the first day and status is given it by the teacher, this will help reduce the stigma factor. Further, by making the tutorial program open to all students as either participants or consumers of the service, the possibility of stigma is again reduced. Some instructors suggest that when first starting a tutorial program tutors who have high esteem or status in the class should be used; the base can be broadened later. Finally, before making a specific assignment for tutorial help, ask the student who is to receive the help if he or she wants this sort of assistance. Tutorial help may be more acceptable to some students in a resource room rather than a technology lab. The involvement of all parties concerned eliminates surprises later.

Integration

Integration in the classroom can occur naturally; at times it can be appropriately assisted. In order to avoid a haphazard approach, the instructor should remain aware of those factors that might influence successful integration, such as peer attitudes, family attitudes, and the student's outlook. The teacher can often avoid unnecessary traumas by consulting with the referring teacher and student. Finally, the teacher must examine his or her own attitudes since the teacher will set the tone for integration. The following activities can make the effort to integrate handicapped students successful from the start.

Involve the Parents

In many instances the parents or guardians have been the sole advocates for the student and they will embrace mainstreaming in the shop eagerly. Other parents might have questions about the transition from specialized programs to less structured ones, or concerns about safety and risk. At the IEP meeting the technology instructor can present his overall program, solicit the insights and suggestions of parents, and confer with the participants on solutions. Another positive step is to carry the involvement of the parents or guardians one step further by inviting them into the class. One can also suggest extensions of the technology curriculum to home and community activities. Such activities might include:

- Attending a science or technology fair at a local museum
- Attending arts and crafts exhibits related to technology (e.g., metal sculpture, wood sculpture, custom car shows)
- Hobbies or kits that can be worked on in the home (e.g., electronics kits)
- School or community courses that relate to technology
- Visiting places that offer postsecondary training programs

These activities could also be assigned to the entire class for supplemental work credit.

Involve the Class

The expanded definition of technology includes expanding our view of man, society, and technology. Besides some of the more obvious topics of futurism, changing working conditions, and leisure outlets, instructors can include topics related to disabled workers and handicapped citizens. Some of these topics could concern the right of handicapped persons to hold jobs without discrimination, services for disabled workers, and rehabilitation engineering. Possible topics for specific discussions include:

- How people become defined by work, leisure, and consumer activities
- How people acquire skills (innate versus training)
- In what ways people are unique and alike
- How a person's abilities and inabilities become a source of cooperation and how a competitive society can promote equal opportunity.
- How discrimination is reflected in the work place and the community
- The meaning of the phrase, "We are all handicapped; some are just better able to disguise their handicaps and play to their strengths."
- The meaning of the phrase "Handicapped people are like everyone else; they're different."
- The right to be different
- Personal attitudes about and experiences with handicapped people or personal inabilities
- How students would want to be treated if they became disabled

Films such as "Try Another Way," by Marc Gold, and "A Different Approach," produced by Norman Lear, combine humor and insight. Films, discussions, and field trips will broaden the understanding of the class and cement the medium upon which integration can flourish. Speakers from different advocacy organizations (see appendix C) and visits to sheltered workshops and training programs will make the experience real and develop the face-to-face encounters that can erase stigma and fear.

Another approach to use is the simulation of impairments. While these are limited in some ways, they can bridge understanding for many learners. Examples of possible simulation exercises are:

1. Hearing impairment: Show a film on technology while students wear ear protection, or turn down the sound. Film and then show a demonstration in which the sound is distorted. Or give a lesson silently that is simultaneously interpreted in sign language.
2. Orthopedic impairment: Have students attempt to get around the technology lab and participate a full day in school while in a wheelchair, on crutches, with one hand tied to their side, or wearing bulky gloves.
3. Visual impairment: Have students take part in a full day at school while wearing blindfolds or blinders to restrict vision.
4. Intellectual impairment: Show a sophisticated film on advanced technology and give a test immediately following it. Give a demonstration with too many details and rules. Make a hand-out with words left out, difficult words, or reversed letters.
5. Emotional impairments: Invite speakers to address the class on mental illness. Present students with a job that cannot be completed in time. Randomly select certain individuals and talk about them with made-up gossip. Discuss the concepts of pressure, stress, and family constellations. Show how different positions within the family, i.e., first-born, determine how one is treated.
6. Feelings of isolation: Ask the class to wear something odd all day long. Assign tutors to part of the class. Select persons randomly to avoid during class.

Follow-up discussions of these types of exercises might include:

- Do you have any relatives with handicaps? Do you see them as different or handicapped first or as people first?
- Has anybody known someone who suffered a permanent injury from an accident? How did it change his/her life?
- What are common slang terms used for persons with disabilities? What are common stereotypes?
- If you were a parent and your child was labeled ____________ or your parents told you your brother was labeled __________, how would you react?
- How would you feel if you always went to a special room in school? If people would not sit with you at lunch or at football games?

- Give examples of personal experiences where you have felt left out or discriminated against.

In actuality, these experiences are shared by most persons in one form or another. *Take A Card, Any Card,* by the author, is a play about handicap awareness. It has been used as an activity by having the students first read the play and then discuss the parts. *Take A Card, Any Card* is now in videotape form for classroom use (available from Martin Kimeldorf). The videotape has been used effectively in training peer tutors.

Examine One's Own Attitude and Practices

In some situations, the greatest barrier to mainstreaming is the teacher's attitude. Put another way, the teacher's perception of his or her own limitations becomes the teacher's perception of the student's disability. Perhaps behind this is the fear of failing to teach a student. It should be realized by now that many techniques relegated to special education are, in fact, standard techniques with slight alterations. Thus, while it does not require a special person to teach a handicapped student, it does take a special teacher to respond to all students with equal enthusiasm or professional commitment—all of whom are special in one way or another.

Often the teacher becomes the model of accepting behavior that the entire class then adopts. Here are some guidelines to consider when teaching or when examining one's current teaching style.

- Don't assume that a speech impediment means an intellectual impairment.
- Don't talk down to students or use "baby talk."
- Avoid raising your volume; it will not change the rate at which someone learns.
- Avoid overprotective attitudes or methods:
  - Don't change your personality around exceptional students; they like to be teased and kidded if that is your normal way of interacting.
  - Allow students to make mistakes; they won't break.
  - Don't spend all your time supervising one particular student or always starting with him when observing; teach independence.
- If instruction does not appear to be successful, ask the student, as well as other instructors, for suggestions.
- Keep the same expectations for behavior that you have for all students:
  - Criticize inappropriate behavior and breaking of rules so the student knows the behavior is unacceptable.
  - Expect the student to work hard, cope with adversity, etc.
- Try not to become self-conscious (e.g., don't worry about using the words "hear," "listen," "see," or "look" around deaf or blind students).

- Demand respect for all students.
- Know the student's name and use it outside of class.
- Don't bring visitors in just to see the special ed students.
- Don't assign peer tutors or other forms of help without first consulting with the student for whom help is suggested.
- Encourage all students to participate in open houses, clubs, crafts shows, social events, and elected positions.

It is often not so much the special assistance offered that the student might be reluctant to accept, but rather how it is offered. When in doubt consult with the referring teacher first and the student next. Obviously, some instances require special preparation, as in the case of field trips. It may be important to check accessibility, at what heights items are visible, whether students can examine things tactilely, if there is excessive noise, or if sign interpreters can accompany the class. Again, talking with the student in question is always wise.

Special consideration must be given the concept of publicity, especially as new programs evolve and naturally accrue a high profile. One disabled student in a class can often become a symbol for all students with special needs. The student can become more isolated if he or she is the one always receiving the publicity, attention from visitors, or getting assistance. Everyone has the right to have his or her confidentiality protected. Students with visible handicaps should always have the prerogative of refusing to be photographed or singled out for extraordinary attention or publicity. Of course, some students welcome the chance to display their abilities or to advocate for handicapped citizens. In either case, it comes back to consultation with the student first.

Paralleling the consideration of publicity is the need everyone has to feel special, that he or she has a contribution to make. It is possible to select certain sequences or activities that allow special needs students to receive attention for what they can do and not for their disabilities. For example, in an electronics class students were first required to "breadboard" their project before making a printed circuit board for final assembly. In one case a student was allowed to move directly to the final assembly stage because the project might not otherwise be completed in time. Soon, the other students had gathered around him, hoping to glimpse what their future project might look like.

Individualizing Instructional Materials and Techniques

The rationale for individualizing instruction is simple: Each learner has a unique way of learning and responding to instruction, and when the instructional materials and methods reflect this, maximum learning takes place. A basic tenet of this outlook is that the educational setting should not be looked upon in Darwinian terms; this setting is not a jungle where the fittest survive or are rewarded most. This does not rule out learning about competition within a positive and structured experience. Nor does this mean that each learner requires a separately designed curriculum. Rather, the degree of individualization required will often correlate with the severity of the handicap. For most learners, if the instructor provides a degree of flexibility in the educational setting, then individual differences can be accommodated. This simple form of individualizing comprises three basic options:

- The option of working at one's own pace or working toward goals within the realm of one's capabilities.
- The option of being evaluated based on one's performance rather than with competitive grading or norm referenced grading.
- The option of receiving one-to-one attention, or modifying instructional materials as needed.

Clearly, all learners could benefit from this type of instruction. A balanced perspective must also accompany these techniques. It should not be translated into techniques that only handicapped students use since this may stigmatize the instructional method. Secondly, socialization skills cannot be acquired if every student works away in an isolated work cubicle, totally alone. In other words, flexibility for all learners that reflects different special needs can become a realistic goal for the entire class. Many teachers already incorporate this teaching style and need only consider some of the additional suggestions contained here.

Teaching Style

Teaching style related to special needs students can be summed up as a process of simplifying methods to attain reasonable goals. The following list is an inventory of techniques to consider:

- Teach only what must be immediately applied (add detail or depth only *after* basic skill has been mastered).
- Keep demonstrations short and with immediately relevant content.
- Teach concepts keyed to short unit goals; attempting to bridge to later instructional units may result in confusion.
  (Teaching the letterpress as raised printing can be one concept. However, relating it to lithography, which takes place on an apparently flat surface and involves additional concepts related to oil and water, may serve to confuse the student.)
- Emphasize experiential learning over lecture/reading format when teaching a new skill.

- Utilize the services of the educational specialist. Provide materials that can be used for practice in the resource room or that can supplement instruction and practice (e.g., safety tests that can be read by other students or the teacher in resource room, micrometers that can be used for practice).
- Provide the option of a further breakdown (task analysis) of the lesson and additional time for mastery of the skill.
- Teach study skills related to the course. For example, when lecturing, provide a sketetal outline for students to fill in or put the major headings on the blackboard first. Teach the SQ3R (scan/survey, question, read/recite/review—consult reading specialist) method of reading for information, and grant credit for those who enlist the technique (e.g., credit for students who supply an outline of the main ideas in a chapter).

Another major method to consider is bringing the reading program vocabulary into closer congruence with the content of the technology program. One way of doing this is to supply the reading specialist with software and hardware that can be used in the reading program. Some examples of these are:

1. Vocabulary lists for tools, materials, and processes plus the real items or pictures from books or magazines. This builds up vocabulary and prepares the student for use of recall skills related to technology.
2. Supply operation sheets, safety lists, or rules that the reading specialist can turn into reading comprehension skills. These skills can build up the student's ability to sequence and recall needed facts.
3. Supply a maintenance chart or symptoms/diagnosis chart that can be related to reading skills. The student begins to use previous knowledge and to build comprehension skills related to inference and logic.
4. Supply different materials for comparison. Many reading and written language programs ask students to "compare and contrast" orally and in writing. Comparing and contrasting different materials can bring added relevance to the assignment (e.g., different wood finishes, different wire sizes, different types of metals or examples of welds, different types of oils, different foundry sands, etc.).
5. Supply examples of concepts to be tested and have the student make up questions for a test.

Teaching Materials—Commercial

Many different training and instructional materials are available today. Some are specifically geared to special needs learners while others are individualized or programmed and can be used by all learners. The disadvantages of commerical products are well known:

1. They often require a large initial capitalization.
2. Materials must be mastered by the teacher and later adapted to the specific requirements of the school curriculum and course methods.
3. Materials are consumed or require some degree of preparation (e.g., engines must be reassembled, tapes rewound, solder replaced).

The advantages are also easily recognized:

1. They have a consistent structure with most of the bugs removed.
2. They can be administered by persons with different professional backgrounds.
3. Inventory and replacement is simplified with a single supplier.

There are four broad categories of materials: vocational evaluation; training materials for special populations; specialized materials for specific disabilities; and training materials, which are self-contained, individualized, and often programmed. Resources and sources for these materials are listed in appendix C.

Vocational Evaluation Materials

Chapter 2 lists several sources of commercial products, along with examples of teacher-made materials. The relationship of assessment to training materials is very close and training materials could, therefore, be developed from already existing evaluation instruments.

Training Materials for Special Populations

Two major types of materials fall into this category. First are the sources of information that can lead the instructor to existing materials that are usually purchased. Examples of these are:

Material Development Center
Stout Vocational Rehabilitation Institute School of Education and Human Services, University of Wisconsin-Stout
Menomonie, WI 54751

National Center for Special Education Materials
University of Southern California, NICSEM, University Park (RAN) 2nd Floor, Los Angeles, CA 90007

The National Center for Research in Vocational Education
Ohio State University, 1960 Kenny Rd, Columbus, Ohio 43210

New Jersey Vocational Technical Curriculum Lab
Rutgers-The State University, 4103 Kilmer Campus, New Brunswick, NJ 08903

These organizations provide bibliographies, summaries, and listings of teaching materials. They represent national networks for collection and dissemination. The other major sources of instructional materials are the companies that supply software (curriculum guides, instructional materials, tests, etc.) and the related hardware (tools, consumable materials, machines, cabinets, etc.). Examples of these are:

- Project Discovery. One big distributor is
  Schoolfutures
  Marylhurst, Education Center
  Box A–O'Hara Rm. 209.
  Marylhurst, OR 97036

This program is geared toward the exploratory level and middle school. Instruction is based on short instructional units using caricatures or cartoon characters.

- Career Related Instruction (CRI)
  Capital Area Career Center
  611 Hagadorn Road
  Mason, MI 48854

These materials are geared toward the entry-level curriculum using a standardized format for vocabulary, tool identification, and processes.

- Vocational Assessment and Development Program.
  Broadhead-Garret Co.
  161 Commerce Circle
  P.O. Box 15528
  Sacramento, CA 95815.

This comprehensive program was developed by one of the traditional suppliers of materials for industrial arts curricula, Broadhead-Garret. It provides instructional guides, hardware, and assessment formats that span vocational evaluation, entry classes, and job preparation programs.

Specialized Materials

These are materials that help compensate for particular disabling conditions. Some are related to one particular impairment while others are for special needs learners in general. Examples of these are:

- Perkins School for the Blind
  Watertown, MA 02172 (no street address given)

This school can supply brailled rulers, compasses, and protractors, and drawing kits that produce raised line drawings.

- Telex
  9600 Aldriche Avenue
  South Minneapolis, MN 55420

  This company produces devices for broadcasting and later amplifying sound (hearing aids) that could be useful for students with hearing impairments in a noisy work setting.
- Kurzweil Computer Products
  264 Third Street
  Cambridge, MA 02142

  This company produces a machine that can decode (read) printed texts and translate them into synthesized speech. This could assist visually impaired as well as learning disabled students.
- Gestetner.

  For address, consult distributor of office equipment or printing equipment.

  This company produces the 319 offset press, which is supposed to be modified to accommodate all learners. The height can be adjusted, dials are brailled and use symbols instead of words, extra guides and shields have been added, levers have been enlarged, and several auditory signaling devices come with visual indicators as well.

The interested reader should consult chapter 3 for listings of additional resources.

Training Materials

Several materials listed in appendix C under "Instructional Programs" are conventional vocational training programs available commercially. These include Ken Cook (welding and small engines); Ideal Developmental Labs (building and home repair); and Basic Skill (by Singer).

Teaching Techniques and Teacher-Made Materials

Many of the materials and techniques listed here may already be familiar to the reader. This is a listing of recognized and successful practices that require only slight modification.

Written Materials

Written materials often used by technology educators include job sheets, information sheets, operation sheets, safety sheets, and worksheets guiding specific assignments. For the student with limited reading ability these may become an impediment to learning or participation. With little advance planning these written materials can be simplified and made more consistent and more readable, thus making them accessible to a broader base of learners. The following suggestions can be used as guidelines in the design of written materials.

1. Simplify the writing style and be consistent:
   Avoid using synonyms (e.g., circuit board for printed circuit board).
   Use short, uncomplicated sentences.
   Use short paragraphs that contain only one major idea.
   List formats when possible, rather than presenting them in narrative form.
   Avoid generalities, unnecessary vocabulary, or extraneous references.
   Use main headings and subheadings for easy access to information.
   Double space.
   Use clear, high quality reproductions.
2. Supply illustrations when possible. Cabinet drawings are preferred to multiviews when first picturing objects.
3. Incorporate checkpoints so that student progress can be monitored. Checkpoints usually require that the teacher observe certain prerequisite behavior before the student continues (e.g., safety test, filling out worksheet, completion of previous steps, or supervision required for given step).
4. Integrate safety into the written directions.
5. Provide vocabulary instruction prior to presenting written material or integrate instruction into the material. Keep nontechnical vocabulary at third-grade level when feasible.
6. Start with simplified worksheets that gradually become more difficult or that lead to other supplemental worksheets. Beginning worksheets should not contain more than ten items to be filled in or read in order to build success into initial assignments. A standard worksheet related to tools, materials, and processes can be broken down as follows:
   Worksheet #1: Name, materials used, project title, sketch. (Draw one of three views. Student is helped with other two views.)
   Worksheet #2: To the list of materials add dimensions, quantity, and costs.
   Worksheet #3: List tools and machines used in fabrication.
   Worksheet #4: List sequence of operations for fabrication.
   Worksheet #5: Write safety rules for items in worksheets #3 and #4.
   Worksheet #6: List finishing steps.

Each worksheet is presented as the student needs it rather than all at once.

7. Mastery of symbols used in technology (blueprints, codes, schematic symbols) can be introduced in a gradual way. Use of the following sequence ensures mastery.

Worksheet #1: Symbol is used with picture or description.
Worksheet #2: Symbol is given in dashed lines that the student must trace; student then adds the name or description.
Worksheet #3: Symbols are used alone (reference sheet is provided; see Westling 1979 for similar idea).

8. Use a handout similar to the entry-level forms in which words can be matched to dashed lines (see figures 5–15 through 5–18).
9. Use transparencies of worksheets to instruct students on their use before handing out worksheets.
10. Provide alternatives within a written assignment: Tapes or tutors can supplement the worksheet. Use performance tests as alternatives for written tests; students are checked off on a written form after adequate performance.

Utilize All the Senses in Conveying Information

When attempting to instruct students about different types of materials, finishes, lubrications, drawings, and qualities, use as many of the senses as possible to accelerate learning. Match drawings to models (kinesthetic-visual), to wood samples (tactile - olfactory - visual), to metal types (tactile - visual: weight, spark test, finish), to tool functions (tactile-auditory: movement, sounds). These are ways of broadening instructional channels. In the ZVI Tactile/Sensory Experiential Learning Module students can examine three-dimensional models related to different wood working practices and concepts. There are examples of different types and patterns of wood grain, sandpaper and related finishes, joints, fasteners, etc. These all fit into a cabinet of small boxes. When the student has a question he looks up the topic, removes a box with the model, and can follow the written information associated with it. Other typical examples include:

- Cutaway functional models for demonstrating functions and parts, which the student can use in the first assignment
- Color-coded parts to assist in assembly and disassembly
- Color-coded views to assist in drawing multiviews
- Color-coded machines keyed to a chromatic sequence to assist the student in recalling the sequence of setup or operation

Provide Alternatives for Drawing or Interpreting Drawings

These alternatives can help the student master drafting skills as well as the use of plan sheets. Examples of these could include:

- Use a fixed inventory of projects. Student must then match the drawing to the project he or she would like to build.

- Provide overhead illustrations of working drawings. These can then be enlarged to the size of the final product and the student can trace them to generate a plan (see Ross 1976).
- Use models instead of blueprints or photo essays instead of operation sheets.
- Provide parts of a project in a prefabricated form that the student must then assemble according to a plan sheet.

Provide Alternatives to Math and Measuring

Many times the requirements of a project assume a mastery of basic math or measuring skills. Often routine use of these skills can be simplified by using some of the following.

- Gauges: go and no-go gauges for drill size, and similar gauges.
- Models and story boards for measuring, where measurements are taken off the model on a story board and transferred for layout on the material.
- Projects requiring accuracy to only one inch. Later projects may require finer precision as the student masters these and uses increasingly more complex scales and measuring devices.
- Volume and quantity measurements simplified by using pictorials of the amount or color-coded references. Examples that relate to plastics, autos, and ceramics can be found in home economics (see Krinke 1980). The symbol (drawing or color) for ½ teaspoon of resin or hardener can be used to supplement regular instructions.
- Charts that help to simplify calculation skills. Use existing charts (e.g., drill tap charts). Make up charts for specific tasks. For example, one instructor provided a gauge that measured standard metal rod diameters. The associated chart related the diameter for different metals (e.g., brass or steel). The chart further displayed information for finding rpm (speeds) and machine feeds.
- Encouragement to use calculators. Provide computation instructions with pictures of the appropriate calculator keys in the directions.

Utilize Instructional Packages Referenced to Curriculum and the Lab

This is similar to using the instructional packages described under entry-level programs. The difference between these and entry-level packages is the greater depth of skill targeted for student mastery. The packages might be more highly specialized in content and thus more numerous. These packages differ from commercial products because they are referenced to the exisitng equipment and tools in the lab. Examples of how this kind of package can eventually look and how they can become incorporated into an instructional system are found in the ZVI method manuals developed by Charles Ross in the Portland Public Schools, Oregon. The creators of the ZVI system claim the following:

- It serves as a quick instructional reference material for all learners
- When students use the references the teacher is freed from spending time on answering minor instructional questions regarding machine operation, procedures, and tool setup
- It assists the teacher in evaluation
- It supports the efforts of team teaching when some of the team members may not be wholly familiar with technical processes

Samples from the instructional manuals developed in the ZVI system for use in wood technology instruction are presented in figures 5–27 through 5–29. Figures 5–28 and 5–8 are from instructional manuals while figure 5–29 is from a test manual. In designing these types of manuals, certain instructional and design principles should be adhered to:

1. Sequential photo essays related to sequence of machine setup, use, and safety are used rather than conventional instruction that supplies most of the information in narrative form with only one or two pictures.
2. Photos should be of actual equipment and tools found in the class and captioned with a single sentence or phrase of relevant information. Arrows and colors can help illuminate key parts in photographs.
3. Manuals should have a pictorial index so the student can immediately determine the usefulness of the particular manual and reference information.
4. Beginning manuals with vocabulary lists associated with pictures are useful.
5. Manuals should be used next to the machines, with provision for easy storage and use (e.g., stands).
6. Similar pictures should be used in the test portion of the system as are used in the instructional part. The test manual can be designed on a pre/post test basis.

In using the ZVI system the student first consults a manual and receives initial supervision or demonstration as needed. Next the student must pass a pretest related to safety and knowledge before being signed off. After this the student can use the equipment independently. In this latter stage the manual is used for a quick reference in future applications. The author and creator claims that his teaching stress was reduced because he was able to serve more students in the woods program and the students appeared more highly motivated because instruction had been simplified and made more accessible. Likewise, Mr. Ross argues that a student aid or peer tutor who is unfamiliar with technical processes could use the instructional packages like lesson plans.

CROSSCUTTING-SHORT PIECES

1 Clamp block on miter gauge.

2 Keep fingers well back from blade.

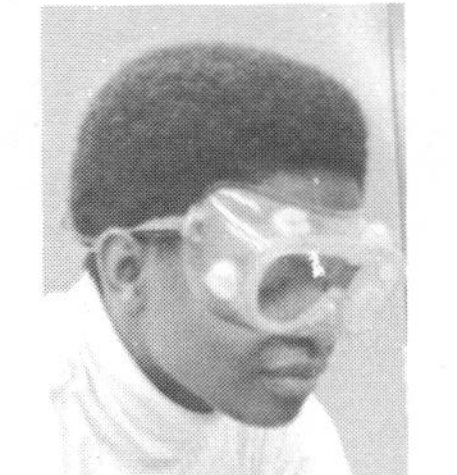

WEAR GOGGLES

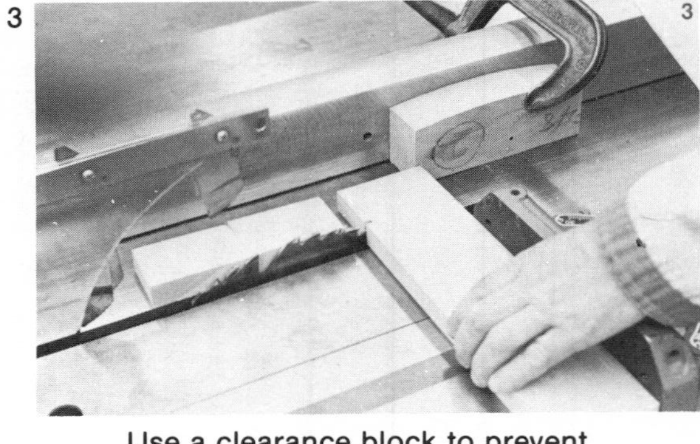

3 Use a clearance block to prevent kickback.

4

White glue
fixes almost
everything
but not
this.
J.B.

Figure 5–27. *(Courtesy of Charles Ross.)*

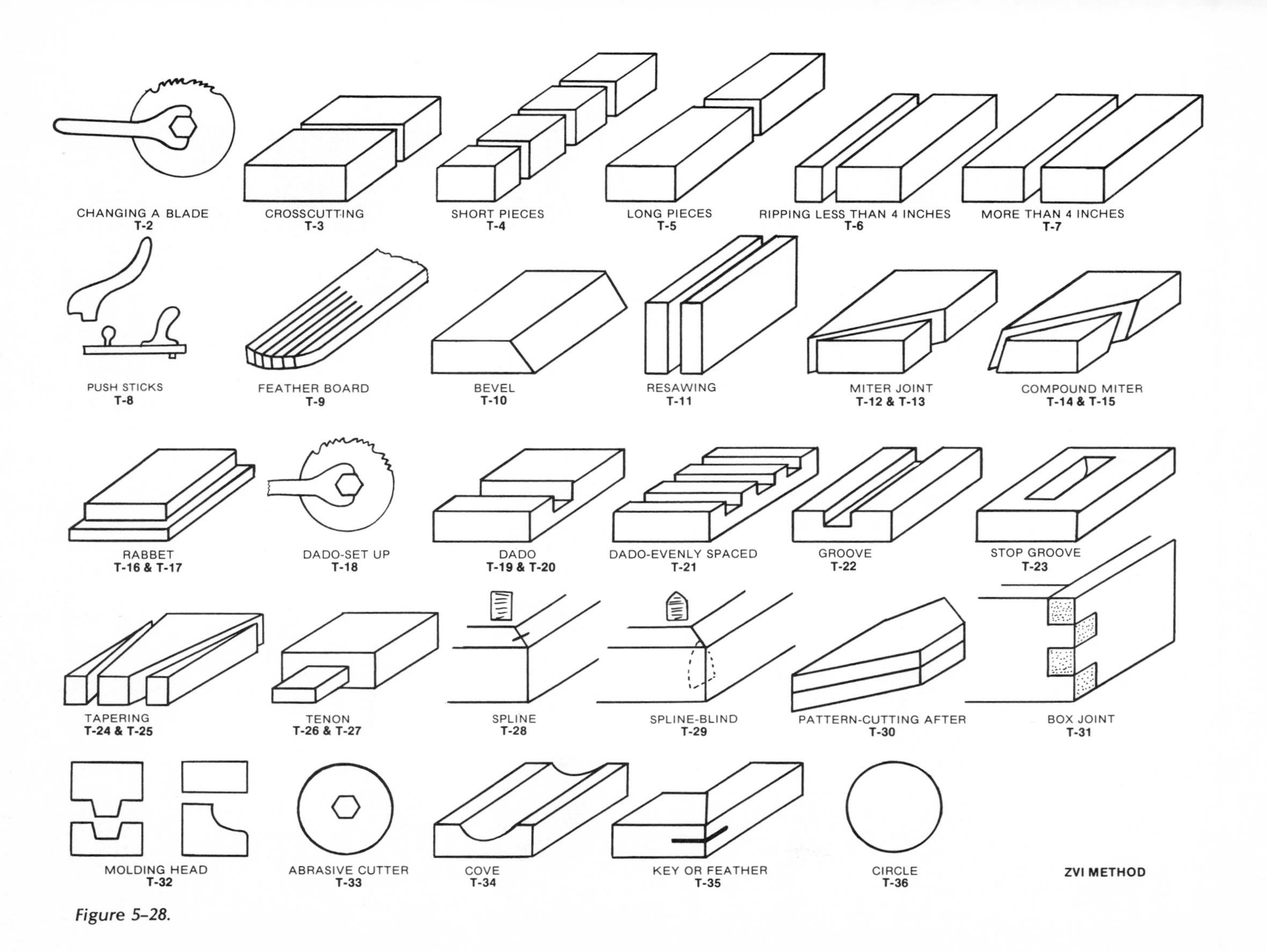
CHANGING A BLADE
T-2
CROSSCUTTING
T-3
SHORT PIECES
T-4
LONG PIECES
T-5
RIPPING LESS THAN 4 INCHES
T-6
MORE THAN 4 INCHES
T-7
PUSH STICKS
T-8
FEATHER BOARD
T-9
BEVEL
T-10
RESAWING
T-11
MITER JOINT
T-12 & T-13
COMPOUND MITER
T-14 & T-15
RABBET
T-16 & T-17
DADO-SET UP
T-18
DADO
T-19 & T-20
DADO-EVENLY SPACED
T-21
GROOVE
T-22
STOP GROOVE
T-23
TAPERING
T-24 & T-25
TENON
T-26 & T-27
SPLINE
T-28
SPLINE-BLIND
T-29
PATTERN-CUTTING AFTER
T-30
BOX JOINT
T-31
MOLDING HEAD
T-32
ABRASIVE CUTTER
T-33
COVE
T-34
KEY OR FEATHER
T-35
CIRCLE
T-36
ZVI METHOD

Figure 5–28.

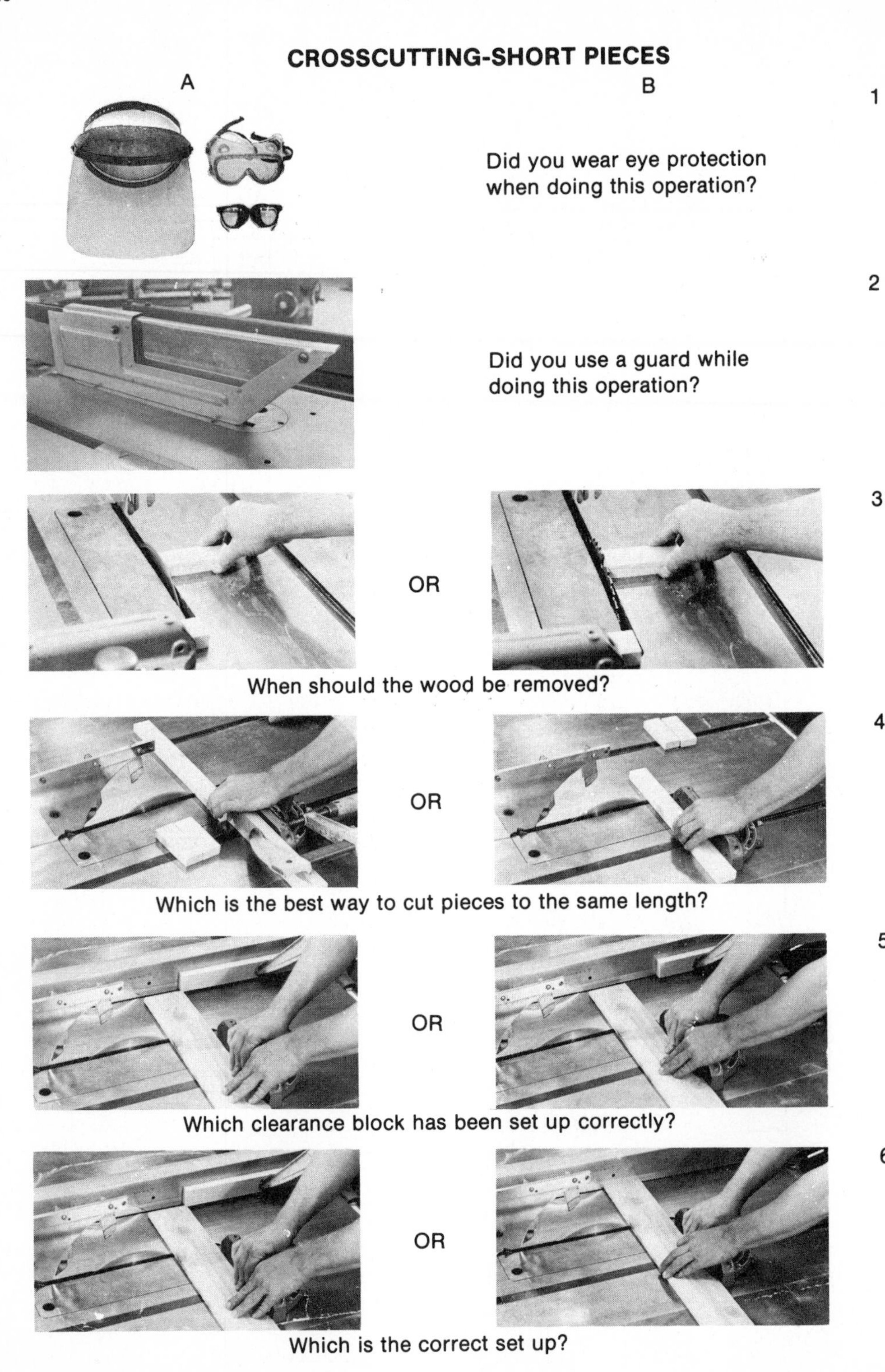

CROSSCUTTING-SHORT PIECES
A
B
1
YES
NO
A
B
Did you wear eye protection when doing this operation?
2
YES
NO
A
B
Did you use a guard while doing this operation?
3
YES
NO
A
B
OR
When should the wood be removed?
4
A
B
OR
Which is the best way to cut pieces to the same length?
5
A
B
OR
Which clearance block has been set up correctly?
6
A
B
C
D
OR
Which is the correct set up?

Figure 5–29. Crosscutting: short pieces (Courtesy of Charles Ross.)

Computer Assisted Video Instruction (CAVI) is another alternative to written instruction. The instructor videotapes demonstrations on safety and set-up, using the current lab equipment and materials. The videotape then is coupled to a written or oral script. After each demonstration the student is given a quiz. If the student does not pass the computer analysis for that part of the videotape, the student needs repetition or more detailed information.

In addition to technology skills, instructors who use CAVI need skills in programming and script writing. A more complete description can be found in the Fall, 1983, issue of the Special Needs *Insider* supplement to the American Vocational Association's journal, *VocEd*, published by AVA, Arlington, Virginia.

Safety Instruction

Concern about safety has become a large obstacle to mainstreaming, when it need not be. Safety is a major concern for technology teachers because they are responsible for their students' safety and are subsequently liable. However, this concern often becomes unreasonably enlarged because of the stigma and prevailing myths regarding students with handicaps. This situation can be brought into a more realistic focus by first stating the record regarding accidents. It is common knowledge that 90 percent of all accidents are caused not by unsafe conditions (for which the instructor is directly responsible) but by unsafe acts. It is also commonly known that unsafe acts are caused not by ignorance or inability but primarily by unsafe attitudes. Unsafe attitudes are exemplified by such activities as horseplay, lack of attention or concentration, boredom, and bypassing required procedures.

The author can state from the personal experience of working with all degrees and types of students with handicaps that being disabled does not make anyone more or less safe in practice or attitude. That is, the same distribution of unsafe attitudes is found in handicapped and nonhandicapped students alike. In fact, the author has never recorded an accident or injury for a student who was handicapped. Likewise, if attitude is a reflection of motivation and cooperation, then the case studies in this chapter suggest that a safer attitude may be found in the entry-level class than in some other teaching labs. This is supported by the fact that nationally, handicapped workers have a better-than-average work and safety record as noted by the President's Committee on Employment of the Handicapped (Razeghi and Halloran 1978). This may possibly suggest that safety concerns for special needs learners and subsequent safety programs (or instruction) can be very similar to standard practices in terms of content and goals.

Finally, there are the legal implications of attempting to exclude a student from a program based on a preconceived notion that handicapped students are inherently unsafe. This would constitute discrimination. This does not mean that any individual student who fails to comply to safety regulations and attend to safety instruction cannot be dismissed from the class because he or she is handicapped (or not handicapped). As guidelines the following are suggestions for consideration:

- Any student may be removed for unsafe conduct. The teacher should attempt to document instruction and subsequent student misconduct.
- If an instructor feels he or she has a potentially dangerous situation because the class contains too many students, this must be put in writing and submitted to the school administrators. If an accident then occurred it would most likely make the administrators responsible.
- A teacher can request consultation and training if he or she feels a lack of appropriate preparation for instructing a student. However, this cannot be used as a basis for excluding a student.

Thus, the requirement for the instructor to take reasonable and prudent measures to ensure student safety remains. In addition, an instructor may wish to consider any of the following techniques to add to the repertoire of existing safety instruction:

1. Provide an entry-level program with a strong safety component.
2. Enlist the support of the education specialist in administering or rewriting written safety tests.
3. Provide the student with peer tutors as outlined previously, which includes documentation of instruction for safety practices.
4. Provide an instructional unit relating to responding to an emergency, accident, stress, or teasing. This might include reporting and exiting procedures, emergency shutdown procedures, first aid related to basic injuries, and related short anecdotal stories that the student must role play or write answers to in the resource room.
5. Post safety rules in large, easy-to-read charts near the appropriate equipment or work area. This can be incorporated into daily instruction for pupils who would benefit from this type of drill.
6. Use standard safety practices and extend them in creative ways: Use Color. Standard color codes exist for coloring machines and environments.

   Use color creatively (e.g., color all machines requiring initial supervision red; color certain parts of machine setup requiring a check by the instructor yellow; color machines requiring permission green; color the path the material travels on and color guards that must be in place blue). (See Schilit and Pace 1978.)

   Supervise initial student attempts. Never accept from any learner "I know how it's done" or "I heard you" to indicate that the student is ready to assume total independence in the lab. For example, many deaf students pick up key words and assume that they have comprehended the total meaning. Another student may mean that he/she paid attention or has seen a parent use the equipment before (rather than actually having done it).

Use films for instruction on safety. See appendix C for a sample list of available resources.

Individualizing Grading/Credits

Individualizing the manner in which a student is graded or receives credit is the final option that represents individualization within the curriculum. Flexibility in this arena can take different forms without watering down the course requirements. A simple example of this is to base a student's grade on attendance, attitude, work habits, and his continued effort to master course materials. In this instance, individual differences will be reflected not so much in letter grades but rather competencies mastered, which can accompany the letter grade. This is a standard technique in many schools, and it has added advantages for the handicapped student. For example, it is common knowledge that in many work experience programs students usually can master the job skills required but often will experience their greatest difficulty in personal adjustment or work habits. This issue can be addressed by using methods that grant credit or grades based both on skill mastery and on work habits/personal behavior adjustment. Coupled with a program that evaluates skill mastery within the range of the individual ability, the product is an individualized grading system. Other options to consider in this domain are:

1. *Pass/Fail.* This system allows greater flexibility for assessing the student's performance (e.g., considering work habits and effort equal to proficiency). It is similar to shaping discussed
2. *Sliding scale of credit.* In this example the amount of credit a student receives is tied proportionately to the percentage of class requirements mastered. To round out this option, the student should have the opportunity to repeat the course for achieving total credit or extend participation to an extra period. The grade in this instance does not reflect the amount of credit the student receives.
3. *Weighted grading system.* This grading system recognizes individual needs for growth. It allows the student to emphasize either work habits/personal behaviors or skill mastery. Either of these two categories can determine the final grade in different amounts. For example, a grade of A might be based on 85 percent skill mastery (test and demonstration) and 15 percent work habits (assessed by peers or teacher or credit earned in personnel systems). For a grade of C a student needs 50 percent skill mastery and 50 percent mastery of work habits. Another approach allows some students to have an 85/15

formula while others would have a 50/50 proportion, depending on individual reasons for placement or enrollment.

4. *Self-improvement.* This approach is adapted from a method used in business education typing courses where it is easy to quantify and measure performance. First, a baseline or pretest is used to establish starting goals. After instruction, the student takes a post-test. The percentage of growth of increase becomes the basis for grading. A variation of this is to establish units of time associated with different tasks (as in automechanics) and then adapt the scale based on an entry-level test that pinpoints goals to achieve.
5. *Self-monitoring.* This technique is not for all students, but its success is well documented, especially as related to work habits and personal adjustment. This method allows the student to assess his/her own performance. Often the teacher establishes a baseline independently or with the student regarding a specific behavior or skill. Later the student takes over the monitoring. It shows a basic trust in the student and suggests to the student that he or she is ultimately responsible for performance.

The grading and credit method chosen should reflect the purpose for which the student was placed in the class.

POSTSECONDARY EDUCATION OPTIONS

Graduating from high school is a life crisis in most students' lives. It is particularly acute for many handicapped learners who suddenly realize that the support and stimulation of their high school program has suddenly evaporated. Three typical options exist for maintaining or continuing their growth: specialized vocational training, adult and college education, and employment. However, these typical options are not always accessible to handicapped citizens due to lack of supportive programs and responsive agencies. Similar to the school predicament, related needs like residential options (ranging from supervised to independent), mobility needs, and health needs must also be considered. If the graduating student exits a life-centered curriculum, then the thrust toward normalization can carry over into the community. This section will mention exemplary programs that represent successful support services designed for young adults who have left the school system. Not only are they exemplary, they are few in number. This fact further underlines the significance of the life crisis mentioned earlier.

Specialized Vocational Training

The traditional agency for vocational training is the local Division for Vocational Rehabilitation (DVR). This agency generally purchases services and supplies instead of performing them. These include on-the-job training, school tuition, materials and transportation costs, costs for vocational rehabilitation training (e.g., rehabilitation workshop or sheltered employment workshop training). Traditionally, this agency serves persons who have suffered an injury. Only a handful of persons with developmental disabilities (e.g., mental retardation, cerebral palsy, brain damage) are served because the agency is geared to those clients who have some "assurance of vocational potential" (see Brolin 1976). Persons who may require intensive training or supervision in order to become competitively employed are often referred to rehabilitation centers with either DVR funds or other sources of funding. The tendency of relying exclusively on workshop placements is coming under increasing criticism because it maintains isolated and segregated work experiences. This situation reflects the lack of training programs that emphasize community placement and training in accord with the normalization principle.

Some notable experiments have taken place. These emphasize community placement or employment at competitive wages. The training may often take place in a specialized work setting, but it is usually geared to employment in the community. Two examples of these follow.

Specialized Training Program

This program, based at the University of Oregon in Eugene, serves the severely and profoundly handicapped. It attempts to prepare young adults for future vocational options based on an analysis of the need for certain types of employment. The jobs result from an analysis of a technology cluster that has been broken down into the component tasks involving hand tools. Many of the clients who were formerly institutionalized are now performing work in a vocational environment while earning competitive wages based on their labors. This program has also branched out into school and community arenas.

TALL (Training Alternatives for Living and Learning)

This exemplary program, cited in a report of the President's Committee On Mental Retardation (1978), is located in Kansas City, Missouri. It provides vocational evaluation, training, and placement in the community. It serves only severely handicapped persons who have been unable to qualify for full-time employment in regular industry. In addition, the program employs an engineer who works on innovative approaches to adapting equipment or job stations so that people with severe physical disabilities can be placed in the community. The engineering is focused around modifications of wheelchairs and the use of existing and cheap technology (e.g., transducers found in toys), which can assist someone in communicating or controlling a piece of equipment.

Programs Based on Work Experience

These programs have a narrower focus than the vocational training programs and are similar to those found in the school systems. Many programs serve as supplementary job placement services, which DVR may help fund or which exist on private funding sources. Increasingly, the move has been toward serving young adults with more severe disabilities. Some programs provide a variety of services while others tend to focus on job placement exclusively.

National Association for Retarded Citizens

Currently the NARC provides monies to employers who hire a mentally retarded citizen and furnish on-the-job training. The monies are used to pay a percentage of the wages during the training period and decrese as the client becomes more productive. Local ARCs often provide job placement counselors and referral services.

Vocational Careers Progam

This program was based at the Portland State University in Oregon. It provided vocational evaluation, prevocational training in mobility skills, self-help skills, and social-sexual education. It also trained special educators in techniques related to career education. The cadre of trainers spent most of their time developing job sites, training clients how to travel to them, and providing on-the-job training and support as warranted. While the project is currently unfunded, it has produced a training manual for working with moderately and severely handicapped adolescents and adults.

CETA (Comprehensive Employment and Training Act) and JPTA (Job Partnership Training Act)

CETA moved toward serving handicapped youths in creative and innovative ways. One such documented experiment is associated with the Specialized Training Program described earlier. In this program moderately and severely retarded adolescents were paired with peer trainers. Both the handicapped clients and the peer trainers were CETA employees. The peer trainers were taught the methods and techniques that have evolved in the Special Training Program so that they could train their partners, take data, and serve as appropriate role models. This program is described by James Clark and others in *Summer Jobs for Vocational Preparation of Moderately and Severely Retarded Adolescents* (in press).

CETA has now been replaced by JPTA (Job Partnership Training Act). It is unclear what will be accomplished under JTPA on behalf of special needs youth and their training. An excellent source regarding the potential of JTPA to provide funds for youth in the area of vocational training, academic instruction and related career needs is the Youth Practitioners Network, The Center for Human Resources, The Heller Graduate School, Brandeis University, Waltham, MA 02254.

Adult/College and College Education

Adult/college education comprises the broadest outlook in programming. Adult education programs can involve vocational education or training, life-experience education, and general college education. Most of these programs take place on a college campus either within the existing curriculum, with support services, or within a modified curriculum. The concept is similar to the placement options and life-centered programs described in chapters 1 and 2.

A program that serves students within the existing college curriculum is commonly found under "Handicapped Student Services" on many college campuses today. This service usually provides support for tutors, notetakers, readers, counseling, accessibility evaluations, advocacy, personal services, attendants, mobility aids, etc. These services are very useful for students with disabilities that are physical or sensory in nature. Catherine Kelly (1977) has written a comprehensive overview of services that can be provided at a postsecondary area vocational school that may be useful to administrators or persons planning to implement such a service. The following colleges and universities and vocational training schools are listed as models for accessibility and student support:

Clover Park Vocational-Technical Institute, 4500 Steilacoom Boulevard SW, Tacoma, WA 98499. The school provides a special needs instructional staff to assist students with a variety of services. These include tutorials, classroom intervention, personal counseling.

DeAnza College, Physically Limited Program, 21250 Stevens Creek Boulevard, Cupertino, CA 95014. The program provides many supporting services related to a wide range of educational and self-enrichment programs offered at the campus.

Gallaudet College, 7th and Florida Ave. N.E., Washington, DC 20002. This college serves deaf and hearing impaired students in a four-year liberal arts setting.

The Hadley School for the Blind, 700 Elm Street, P.O. Box 299, Winnetka, IL 60093. This is a correspondence school for blind students with more than 4,500 students from every state in the U.S. Tuition is free and the school supplies brailled and recorded correspondence courses. College level courses are often available through cooperating universities in Indiana, Wisconsin, and California.

National Technical Institute for the Deaf (NTID), Rochester Institute of Technology, 1 Lomb Memorial Dr., Rochester, NY 14623. This program offers comprehensive support services for deaf students wishing to participate in a vocational training program. Students are enrolled at the Rochester Institute of Technology (RIT), with NTID serving as a valuable partner on campus.

Southern Illinois University, Specialized Student Services Office, Carbondale, IL 62901. The office provides services to students with visual, learning, hearing and mobility impairments. The emphasis is upon integration in campus life.

University of California, Riverside, Handicapped Student Services, 11321 Library South, Riverside, CA 92521. This campus is considered 95 percent accessible in terms of physical barriers. They provide an array of services including financial aid, personal and mobility resources, and preadmission counseling.

University of North Dakota, Student Opportunity Programs, Grand Forks, ND 58201. This campus has developed a systematic method for determining a student's functional needs in a classroom setting. These are based on lists of special needs referenced to specific disabilities.

Another form of adult education involves a modified curriculum. This might include a regular course that features open entry and exit combined with individualized student programming. An example of such a program is the Food Service Program at Portland Community College in Oregon. Students can often get financial assistance from DVR for tuition and related costs. The student can also take this practical, hands-on course and not be enrolled in a full-time curriculum. Thus, the college is used as a training site. Another example of this would be the programs worked out jointly between the Olympia School District and the Olympia Technical Community College in Washington. In this case, high school students participate in specialized training programs in courses not normally offered by either educational institution: building maintenance, landscaping and grounds maintenance, etc.

Perhaps the most comprehensive example of a modified college curriculum can be found in Denver, Colorado in The College for Living, which has been replicated in more than twenty sites nationally. Often these programs evolve from the joint efforts of local advocacy organizations like the Association for Retarded Citizens and the local community college. It was cited as an exemplary program in the 1978 report of the President's Committee on Mental Retardation. The program serves developmentally disabled adults in a regular college setting and utilizes volunteers as instructors. It provides a rich and broad-based life-centered curriculum with such courses as money management, riding the bus, assertiveness training, general experiences (for severely handicapped students), getting along with others, creative cooking, symphony appreciation, keeping healthy, human sexuality, dating, camping and white-water rafting, and woodworking. The staff has produced a package for dissemination that provides curriculum outlines, general outlines of programming, and materials for training volunteers.

Adult education can also extend into the community through local parks and recreation programs, volunteer organizations, and private facilities. The creative endeavors of the Kansas Association for Retarded Citizens are worth mentioning here. They have developed such things as tennis tours, coffee houses (which also serves to recruit and integrate handicapped citizens into the entertainment arena), rodeos, and art festivals.

SUMMARY

Currently, regular education has difficulty in meeting the needs of most handicapped students, as evidenced by the large numbers of students who leave the school system prematurely or who leave without useful skills. A life-centered educational model is one way of addressing this educational dilemma. The need to develop links and cooperative arrangements between schools and community agencies is vital if the special needs student is to have any realistic chance at benefiting from a life-centered educational experience. This sort of cooperation is seen as a high priority in the U.S. Office of Education (National Association of State Boards of Education 1979). Cutting off services after high school and expecting the student to sort through the maze of existing programs and advocate for needed ones is unrealistic and reduces the value of our investment in secondary programs. In an era that questions the role of technology and proposes constructs like appropriate technology, the subject of appropriate education cannot long be ignored.

EXPLORATIONS

Activities

Attempt to create any of the materials described based on local needs: entry-level curriculum outline; instructional packages; cooperative instructional agreements; curriculum analysis; tutorial system.

Examine current written work in technology curriculum in light of adaptations mentioned.

Visit a postsecondary training site.

Help create a program relevant to an adult education program.

Design a peer tutoring system that includes a handicap awareness section. This may be a joint project with personnel in special education. Consult the following reference as a starting point:

People Just Like You: About Handicaps and Handicapped People (An Activity Guide). Committee on Youth Development, The President's

Committee on Employment of the Handicapped, Washington, DC 20210. (For sale from the Superintendent of Documents, U.S. Government Printing Office, Washington, DC 20402) n.d.)

Readings

Consult any of the references cited.

REFERENCES

Babich, Betsy. "HERO Helps Individualized Instruction Materials Development and Process." *Illinois Teacher*. Vol. XXIV. 203–206, March/April 1980.

Bellamy, Thomas G.; Wilson, Darla J.; Adler, Ellen; and Clarke, James Y. *Specialized Training Program*. Eugene, OR: Center on Human Development, University of Oregon, n.d.

Boland, Sandra. "It's Happening: Voc Educators Teach the Handicapped." *Education Unlimited*. 9–11, April 1979.

Boland, Sandra, "Curricular Adaptions for Mildly Handicapped Secondary Students." *Education Unlimited*. 49–51, Nov. 1979.

Brolin, Donn E. *Vocational Preparation of Retarded Citizens*. Columbus, Ohio: Charles E. Merrill, 1976.

Chandler, Theodore A. "Towards a Less Restrictive Environment: Making the Problem the Solution." *Education Unlimited*. 2:21–22, March 1980.

Clarke, James Y.; Greenwood, Lynn M.; Abramowitz, David B.; and Bellamy, Thomas G. *Summer Jobs for Vocational Preparation of Moderately and Severely Retarded Adolescents*. Eugene, Ore. Special Training Program, Center on Human Development, in press.

Conroy, Michael T. "Instructional Sheets for Students with Reading Difficulties." *Industrial Education*. 68: 32–34 Nov. 1979.

Crowell, Caleb E. *Math For The World of Work* (a workbook). New York: Educational Design, Inc., 47 W. 13th Street, New York 10011.

Dale, Mary E. "Peer-Tutoring: Children Helping Children." *The Exceptional Parent*. 9:E26–E27, August 1979.

Dussault, William, attorney at law. Based on a letter to the author dated April 26, 1979.

Edwards, Jean. *Vocational Careers Training Model: A Training Manual for Those Working with Moderately and Severely Handicapped Adolescents and Adults*. Portland, Ore: Portland State University, 1978.

Holeman, Penny. Taken from notes during student teaching assignment under Ms. Holeman. Portland State University, Portland, Ore. Spring 1979.

Kelly, Catherine. *The Development of Individualized Supportive Services for Physically Limited Adults at a Post-Secondary Area Vocational School*. Waco, Tex.: McLemon Community College, Eric ED 146–345, 1977.

Kimeldorf, Martin. *Take a Card, Any Card*. In *Open Auditions*. Portland: Ednick Communications, P.O. Box 3612, Portland, OR, 97208.

Krinke, Laurel. "Are Home Economics Teachers Ready for Trainable Mentally Handicapped Students in the Classroom?" *Illinois Teacher*. 224–226, March/April 1980.

LaBell, Jan. *Training Unit Manual Outline*. Olympia, Wa: Thurston County Cooperative Social Services, n.d.

Lewis, Eleanore Grater, and Fraser, Kathleen M. "Avoiding Classroom Tokenism." *The Exceptional Parent*. 9:E7–9, Aug. 1979

Michealis, Richard. *Handbook for Graphic Arts Portable Workshop*. North Clackamas, Ore.: North Clackamas School District, EPC Project 77/78–15.

National Association of State Boards of Education. *Vocational Education of Handicapped Youth State of the Art*. Washington, D.C.: NASBE (44 North Capitol Street, N.W., 20001) 1979.

O'Brien, Thomas. "Vocational Training and Placement of Handicapped Students." *Industrial Education*. 69:29–31, March 1980.

Phelps, Alan. *Instructional Development for Special Needs Learners: An In-Service Resource Guide*. Normal, Ill.: Illinois Network of Exemplary Programs for Handicapped and Disadvantaged Students (ISU), 1975.

President's Committee on Mental Retardation. "The Teaming of Regular and Special Teachers." In *Mental Retardation: The Leading Edge—Service Programs That Work*. Washington, D.C.: HEW, Office of Human Development Services, MR 78.

President's Committee on Mental Retardation. "Job Training That Industry Understands: Training Alternatives for Living and Learning (TALL)." In *Mental Retardation: The Leading Edge—Service Programs That Work*. Washington, D.C.: HEW, Office of Human Development Services, MR 78.

President's Committee On Mental Retardation. "Mentally Retarded Persons Go to College." In *Mental Retardation: The Leading Edge—Service Programs That Work*. Washington, D.C.: HEW, Office of Human Development Services, MR 78.

Razeghi, Jane Ann, and Halloran, William D. "A New Picture of Vocational Education for Employment of the Handicapped." *School Shop*. 37:50–53, April 1978.

Ross, Chuck. *ZVI Student Design Aids*. Portland, Ore.: CEDS, 1976.

Ross, Chuck. *ZVI Tactile/Sensory Experiential Learning Module*. Portland, Ore., CEDS, 1979.

Schilit, Jeffrey, and Pace, Thommy J. "High Risk Employment and the Mentally Retarded Student." *Career Development of Exceptional Individuals (CDEI)*. 1:91–103, Fall 1978.

Vocational Education/Special Education Basic Class Task Force *Vocational Education/Special Education Basic Class*. Madison, Wis.: Madison Metropolitan School District, 1979.

Westling, David L. and Koorland, Mark A. "Some Considerations and Tactics for Improving Discrimination Learning." *Teaching Exceptional Children*. 11:97–100, Spring 1979.

Appendix A

TRAINING UNIT MANUAL OUTLINE by Jan LaBell

TRAINING UNIT MANUAL OUTLINE

I. Introduction
   A. Title: "Fine Measuring Unit"
   B. Job or program analysis on which training unit is based
      1. Class or job title: Small Engines 1 & 2, Olympia High School, Olympia, WA
      2. Date analysis done: March 1980
      3. Teacher interviewed: Tim Carlson
      4. Specific course requirements:
         a) Students must show interest in class, assume responsibility
         b) Students should have access to a small engine
         c) Motorcycle engines discouraged
         d) Students will be required to purchase their own materials for projects, although there is no shop fee
   C. Duration: The class is an all-year class according to the instructor; however, it is possible for a student to take the class for one semester.
   D. Purpose of training unit: partial preparation for entry into the small engine class.
      1. To familiarize students with parts of an engine that need fine or critical measurements (measurements to .001 of an inch)
      2. To familiarize students with correct tools used for fine measurement (wire feeler gauge, blade feeler gauge)

3. To have students practice tool use measurement of fine measuring units (measurements to .001 of an inch)
4. To familiarize students with reading fine units of measurements
5. To familiarize students with writing out fine units of measurements
6. To evaluate students on items 1 through 5 as this is an important section of the small engines class.

II. Administration
   A. Supplies in training kit
      1. Spark plugs
      2. Blade type feeler gauge
      3. Wire type feeler gauge
      4. Answer keys
      5. Recording sheets
   B. Student orientation
      1. Class description (from O.H.S. course selection)
         a) Small engines 1 & 2: Full-year course, 2 credits
         b) Elective: 9, 10, 11, 12
         c) Prerequisite: none
         d) Supplies or fees: Engines to repair, parts, coveralls, and oil for engines
         e) Instructional objectives:
            (1) To correctly identify and demonstrate specific safety rules and practices used in the small engine repair business
            (2) To accurately identify, select, and use tools used in small engine repair
            (3) To accurately identify the name, function, and position of every small engine part
            (4) To accurately identify and repair problems on several engines
            (5) To accurately and properly rebuild all major systems in a small engine
      2. Pictures: (see facing page)
      3. Future employment and possible places for employment for working on small engines:
         a) small engine shop
         b) motorcycle shop
         c) lawnmower or small equipment shop
         d) chain saw shop
         e) maintenance department for state, city, private companies, that use equipment requiring small engine maintenance
         f) parts man or counter man

         Mr. Carlson notes that usually one or two students per year do find employment in the small engine repair field. The employment is typically a low position with on-the-job training.

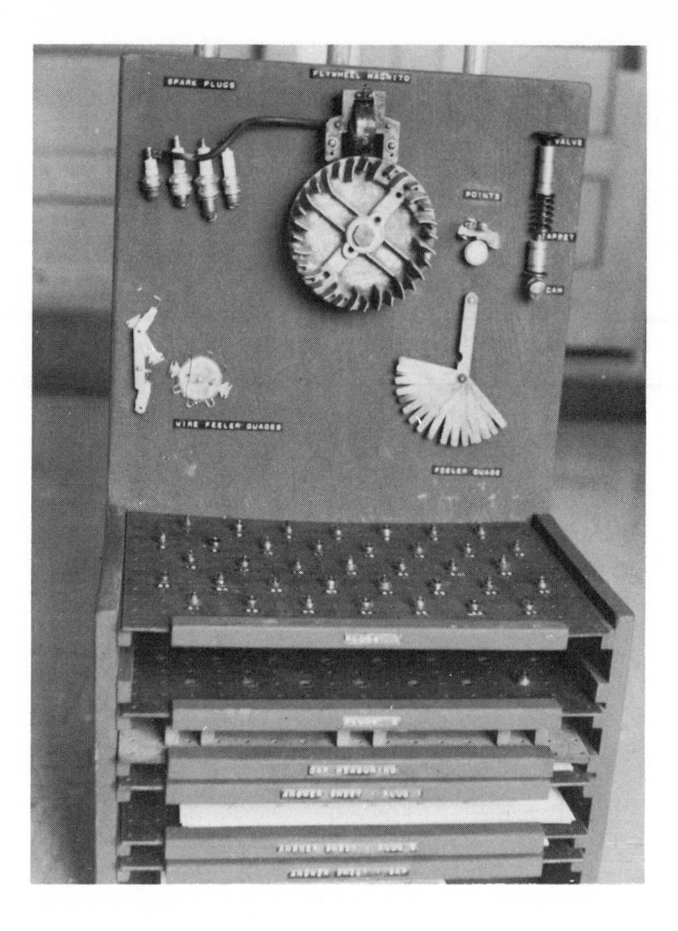

C. Instructions and demonstration

Teaching Display
Fine Measuring

Part I: Spark Plugs

(Text in smaller type is for instructor to read to students.)

The display in front of you is to help you get practice in measuring. You will need to be familiar with this type of measuring when dealing with a small engine or any gas engine. By looking at the display board you can see some areas that need very accurate measurement.

Point out examples to students.

Spark plug gap, point gap, valve lash gap, and flywheel magnito gap all are measured in thousandths of an inch measurements.

To make these measurements you need special measuring tools. To measure spark plug gap you use a wire feeler gauge.

Point out examples.

To measure point gap, valve lash, flywheel magnito gap, you use a feeler gauge.

Point out examples.

Different spark plugs have different measurements. It is important that you know how to measure the spark plugs for an entire engine correctly. For example, some engines require the spark plugs to be set at fifteen thousandths of an inch (.015), while other engines require the spark plugs to be set at thirty thousandths of an inch (.030).

________________________ (name of student), I want you to start by taking out the board marked "Plugs 1" and slide it into the top grooves. Using the wire feeler gauge, measure the distance on the spark plug between the electrodes. This space is called the plug gap. Notice the wire feeler gauge. It has different sizes of wires. Each wire is marked with a number. The wire that slides snugly in the plug gap is the correct one. The numbers marked on the gauge are written in measurements of one thousandths of an inch. If the gap of a plug is twenty-five thousandths, it is written .025.

Show the student on the wire feeler gauge. Show student other examples (.030, .022, etc.).

________________________ (name of student), I want you to start by measuring the gap of these spark plugs.

Point to plugs.

You will record your answers on this sheet.

Hand student the score sheet.

Write the measurement of the plug on the score sheet, like this:

Instructor will do the first one.

When you are finished, take out the drawer marked "ANSWER SHEET—Plugs 1."

You can correct your own paper. Match your answers with the answers on the score sheet. Put the total number of correct answers in this space. If not all answers are correct, ask instructor for help.

Show student the correct place.

If answers are all correct, then go to the drawer marked "Plugs 2." When finished, take out the answer sheet from the drawer marked "Answer Sheet—Plugs 2" and correct your answers.

Part II: Measuring Gap

Other parts of an engine that often need measuring are valve lash (valve gap), point gap, and flywheel magnito gap. All of these measurements are very critical. Different engines and different parts all have different measurements. To measure these parts of an engine correctly you use a blade type feeler gauge.

________________________ (name of student), I want you to practice measuring with a blade type feeler gauge. Start by taking out the board marked "Gap Measuring" Put it on the top groove. Using the feeler gauge, measure the distance between the pegs. Notice the blade type feeler gauge. Each blade is a different size. The sizes are shown in thousandths of an inch. The one you are using measures from one and a half thousandths (.0015) to twenty-five thousandths (.025). The blade that fits snugly between the pegs is the correct one. If the measurement of the gap is eleven thousandths, it is written .011.

Show student the blade feeler gauge. Show other examples (.007, .025, .018).

______________________________ (name of student), I want you to practice measuring the gaps using the blade feeler gauge.

Point to sample gaps.

You will record your answers on this sheet.

Hand student the score sheet.

Write the measurement of the gaps on the score sheet like this:

Instructor will do the first one.

When you are finished, take out the drawer marked "Answer Sheet—Gap Measuring." You can correct your own paper. Match your answers to the answers on the score sheet. Put the total correct in this space.

Show the student the correct place.

You will notice there are three (3) correct answers. The answer in the middle is the most correct. The answer on the left is one thousandth of an inch too tight. The one on the right is one thousandth of an inch too loose. You should try to get the correct "feel" so your answer is the same as the one in the middle.

III. Scoring

Included in the kit are examples of the recording sheets, two for measuring plugs and one for measuring gap (one of each is shown here). Included in the training unit are the respective answer sheets. There are a possible thirty-seven correct answers for measuring plugs. There are a possible twenty-one correct answers for measuring gap.

In order to master measuring with either a wire type feeler gauge or a blade type feeler gauge, 100 percent accuracy is required. Any missed on the score sheet should be remeasured by the student. A student should repeat the process until he or she is capable of measuring the entire sample without any errors or until it is determined that the student does not have the ability to do fine measurements. This would be up to the discretion of the instructor.

IV. Insights

Some points to observe while a student is working on the measuring units include:

1. How does the student handle tools?
   a) Which hand is used?
   b) How is the student's hand control? Shake? Tremor?
   c) Does the student force the feeler gauge?

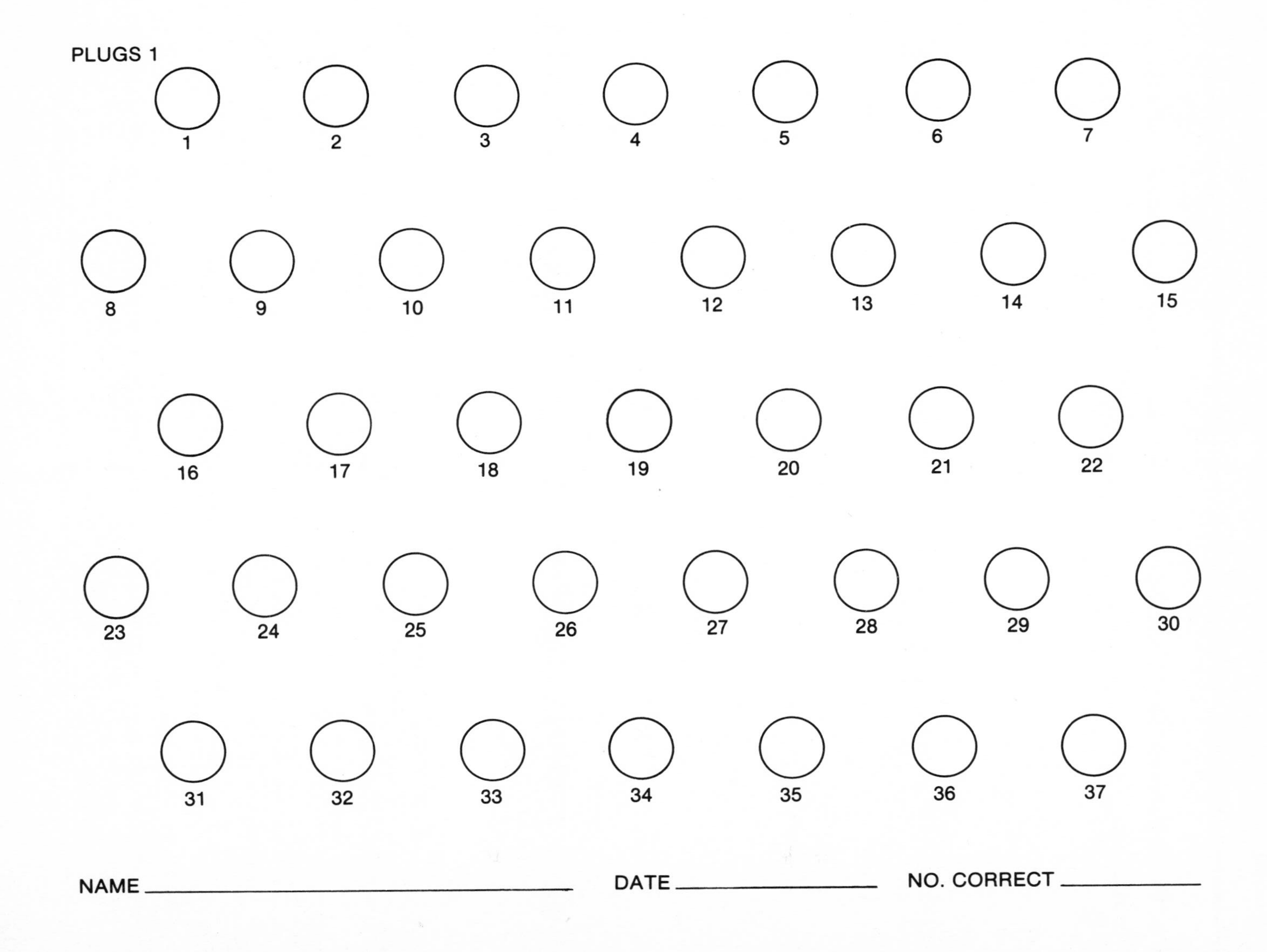
PLUGS 1
1
2
3
4
5
6
7
8
9
10
11
12
13
14
15
16
17
18
19
20
21
22
23
24
25
26
27
28
29
30
31
32
33
34
35
36
37
NAME
DATE
NO. CORRECT

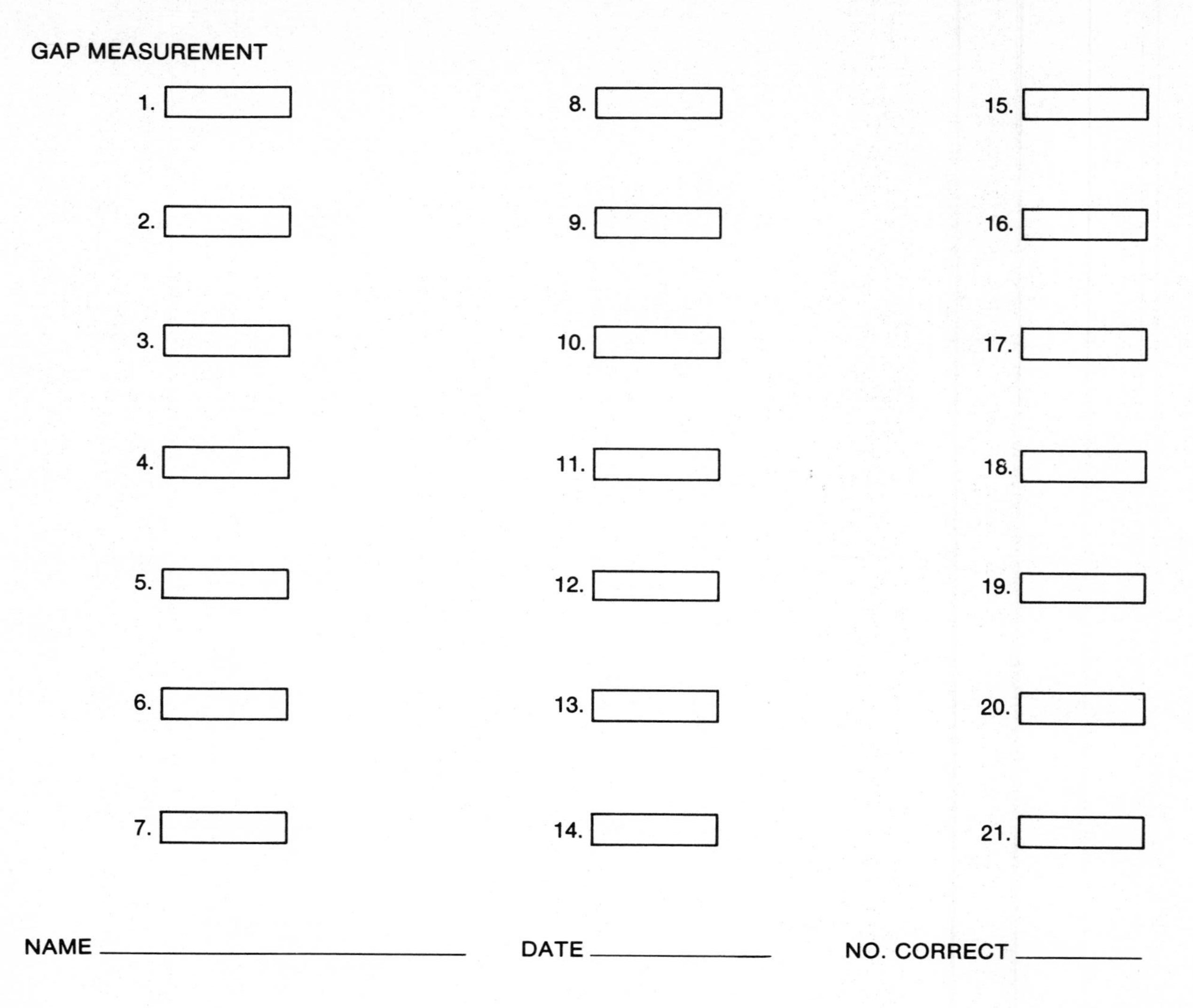
GAP MEASUREMENT
1.
2.
3.
4.
5.
6.
7.
8.
9.
10.
11.
12.
13.
14.
15.
16.
17.
18.
19.
20.
21.
NAME
DATE
NO. CORRECT

   d) Does the student follow a procedure?
   e) Does the student work at an acceptable rate? Fool around?
   f) Does the student play with the display or play with the tools?
2. How does the student record the answers?
   a) Does he/she record the answers neatly and correctly?
   b) Does he/she put the correct answer in the correct space?
   c) Does the student answer the questions honestly?

(The next step after correctly measuring plugs and gap would be to have the student practice setting the gap in plugs, points, etc.)

V. Construction

The entire training unit could be redesigned to fit the purpose of an individual teacher. The important point is that the unit allows the student to work at his or her own rate. The unit allows the student to practice measuring, see his or her errors, and to correct these errors.

Most of the materials for the unit can be obtained from a small engines class. The only cost incurred was for the feeler gauges (wire gauge, $1.50; blade type feeler gauge, $3.00).

Included in the manual are photos of the training unit.

Appendix B

SMALL GROUP EXPLORATORY INSTRUCTIONAL MODULES

The following worksheets and instruction plans represent examples of possible models that could be duplicated for instruction. The worksheets illustrate the main focus of the lesson and pertain to the hands-on components. The instructional guide is similar to an outline for instruction designed to be given in small-group talk and demonstration format. The projects cover:

Industrial mechanics
Electricity/Electronics
Metals: welding
Metals: sheet metal
Graphics: photography
Graphics: drafting

Projects adaptable to woods can be found in chapter 5 in the discussion of adapted exploratory class projects. Chapter 5 also contains pictures of sheet metal projects and equipment used in photography related to the examples in this appendix.

The examples contained in Appendix B are based largely on the prevocational program created by Richard Michealis, a special education instructor in the North Clackamas (Oregon) School District. He has created many other projects spanning all current technology clusters. The instructional units were written by Martin Kimeldorf (electricity/electronics is not from Mr. Michealis's program).

The illustrations were drawn by Michael McHale and Todd Flynn. Both artists illustrated these materials while attending Olympia High School's drafting program. The materials for the illustrations are based primarily on the hand-outs used in Mr. Michealis's program as well as on material used in the author's program at Washington High School in Portland, Oregon.

Technology: Industrial Mechanics

Goal	Explore automobile ignition
Product	Ability to set ignition points
Materials	Mock-up of point set, including rubbing block Feeler Gauge Screwdriver Hex wrench
Lesson	
Concepts	Parts identification of ignition system (Student fills out Worksheet 1) Spark Ignition related to point gap
Demonstration	Opening and closing of points Spark plug and coil (voltage step-up) Use of feeler gauges Adjusting point set gap

WORKSHEET 1

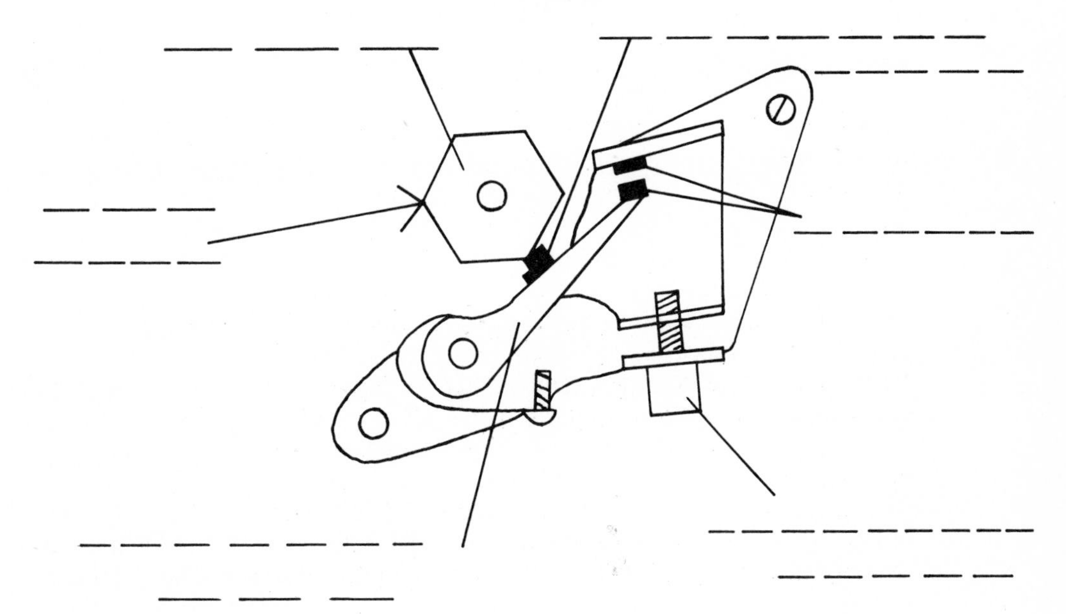

NOTICE THE GAP BETWEEN POINTS
MATCH THE WORD TO THE CORRECT PART

WORD LIST: CAM POINTS BREAKER ARM ADJUSTING SCREW RUBBING BLOCK CAM LOBE

Technology: Metal

Goal	Explore welding apparatus
Product	Knowledge of welding environment and apparatus
Materials	Films on arc and gas welding, welding safety
	Welding torches, arc holders, rods and electrodes
	Protective welding equipment (eye/face protection, gloves, apron)
Lesson	
Concepts	Identify materials and processes associated with welding (student fills out Worksheets 2 and 3)
	Discuss hazards of environment (student watches safety film, puts on protective equipment)
	Discuss content of welding course (student watches welding films)
Demonstrations	How to adjust regulators for gas welding and current for arc
	How torch, filler rod, and electrode holder are manipulated (speed, angles, heights)
	Student goes to welding lab for two periods to observe and experience welding processes with hand-over-hand instruction as necessary

WORKSHEET 2

WRITE THE CORRECT NAME OF THESE PARTS IN THE SPACES BELOW

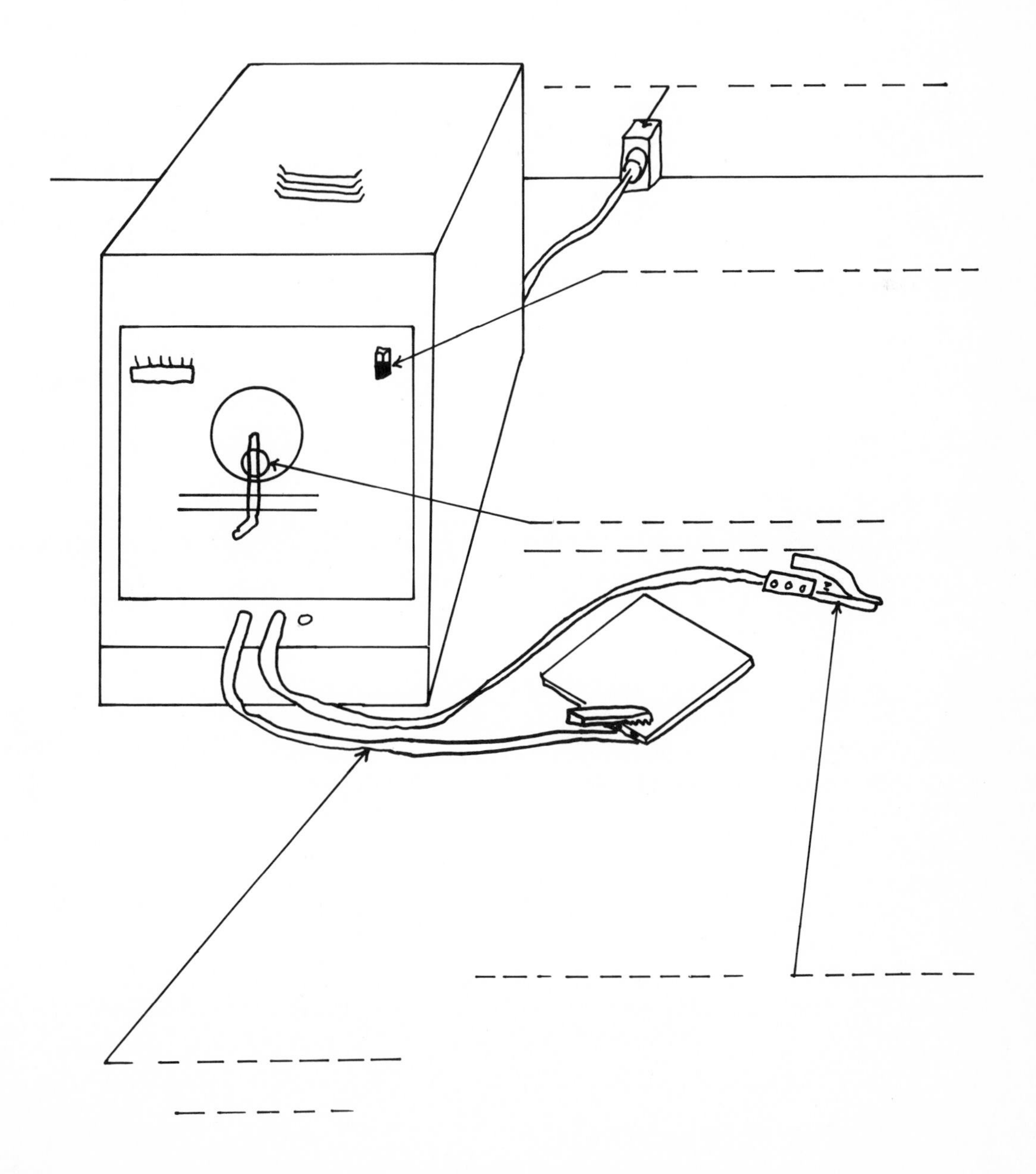

WORKSHEET 3

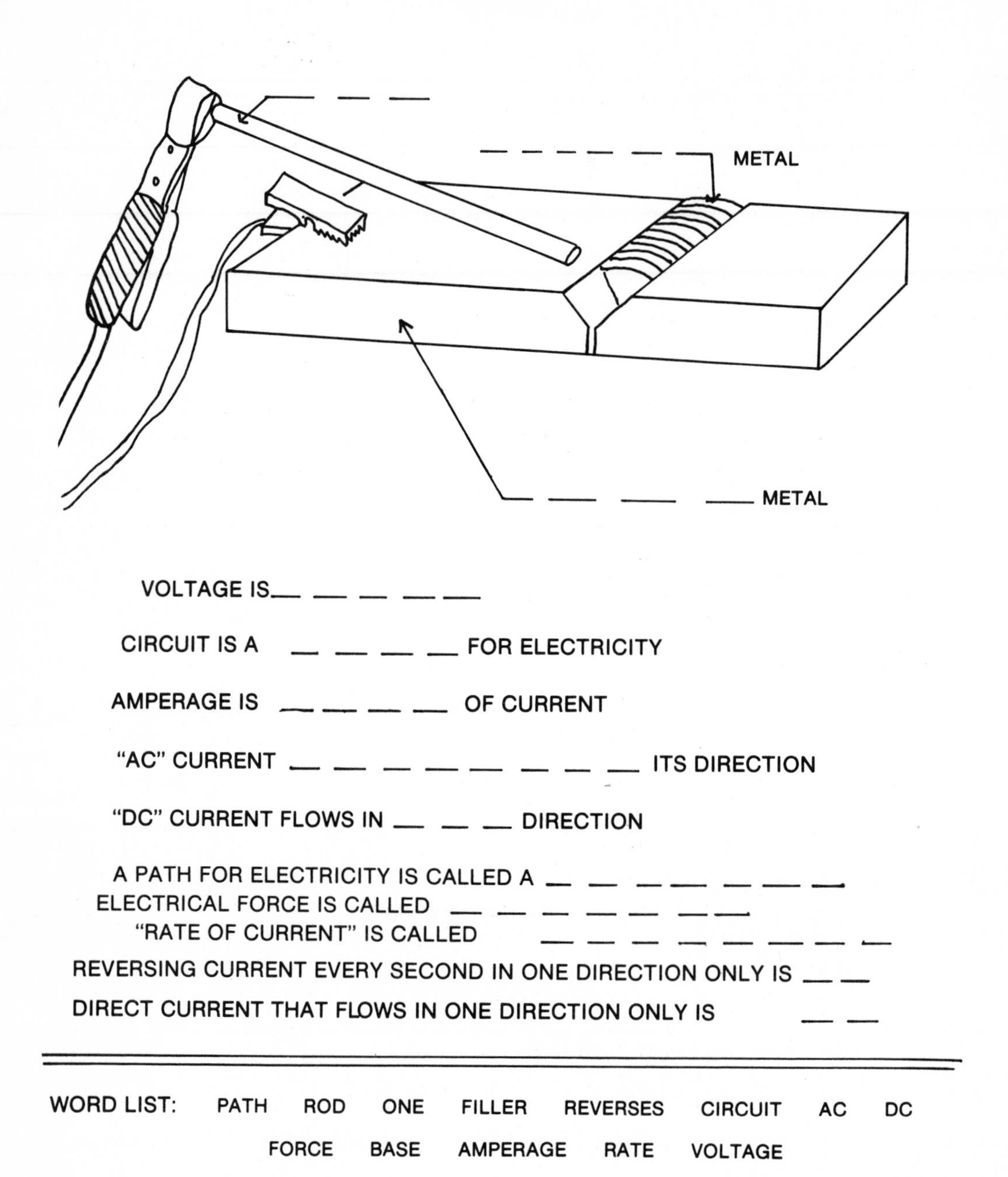

VOLTAGE IS_ _ _ _ _

CIRCUIT IS A _ _ _ _ FOR ELECTRICITY

AMPERAGE IS _ _ _ _ OF CURRENT

"AC" CURRENT _ _ _ _ _ _ _ _ ITS DIRECTION

"DC" CURRENT FLOWS IN _ _ _ DIRECTION

A PATH FOR ELECTRICITY IS CALLED A _ _ _ _ _ _ _

ELECTRICAL FORCE IS CALLED _ _ _ _ _ _ _

"RATE OF CURRENT" IS CALLED _ _ _ _ _ _ _ _

REVERSING CURRENT EVERY SECOND IN ONE DIRECTION ONLY IS _ _

DIRECT CURRENT THAT FLOWS IN ONE DIRECTION ONLY IS _ _

WORD LIST: PATH ROD ONE FILLER REVERSES CIRCUIT AC DC FORCE BASE AMPERAGE RATE VOLTAGE

Technology: Electronics/Electricity

Goal — Explore field of electronics/electricity and learn basic skills

Product — Skills associated with soldering components and repair of light bulb socket

Materials

Electronics

- schematic
- components
- soldering iron and stand
- wire cutter/stripper
- alligator clip
- circuit board
- non-burn surface
- safety glasses

Electricity

- light bulb socket
- plug and cord cap
- light bulb
- safety glasses
- wire stripper
- house wire #12 or #14
- ohmmeter (continuity tester)
- screwdriver

Lesson

Electricity

Concepts and Demonstrations:
- Current, voltage, watts, and power
- Safety around electricity
- Continuity testing of circuit
- Disassembly of light bulb socket
- Stripping wire
- Underwriter's knot
- Wrapping wire around screw (clockwise)
- Filling out a worksheet (see Worksheet 4)

Electronics

- Electronics field and its relation to communication
- View film on field of electronics
- Components: pictorials and schematics
- Chassis and printed circuit boards
- Use of projects and kits as leisure activities
- Soldering: wire stripping
  - safe use of soldering iron
  - first aid for burns
  - good and bad soldering
- Filling out a worksheet (see Worksheet 5)

Technology: Metals

Goal — Explore Sheet Metal Work

Product — Napkin holder, candy dish (box), picture frame holder (see chapter 5)

Materials
- Ferrule maker (see chapter 5 for example of one in operation)
- Sheet metal and tin snips (roof flashing or offset press discards)
- Patterns and scratch awl
- Bar folder (see chapter 5 for portable bar folder)
- Scissors, tape, straight edge

Lesson

Concepts
- Working from a two-dimensional plan student traces pattern from worksheet on one-inch square graph paper (see Worksheet 6). Then student cuts out pattern and folds it up.
- Safety in use of sheet metal tools and materials

Demonstration
- Use of the ferrule maker
- Use of patterns and scratch awl for tracing
- Demonstration of fabricating projects:

WORKSHEET 4

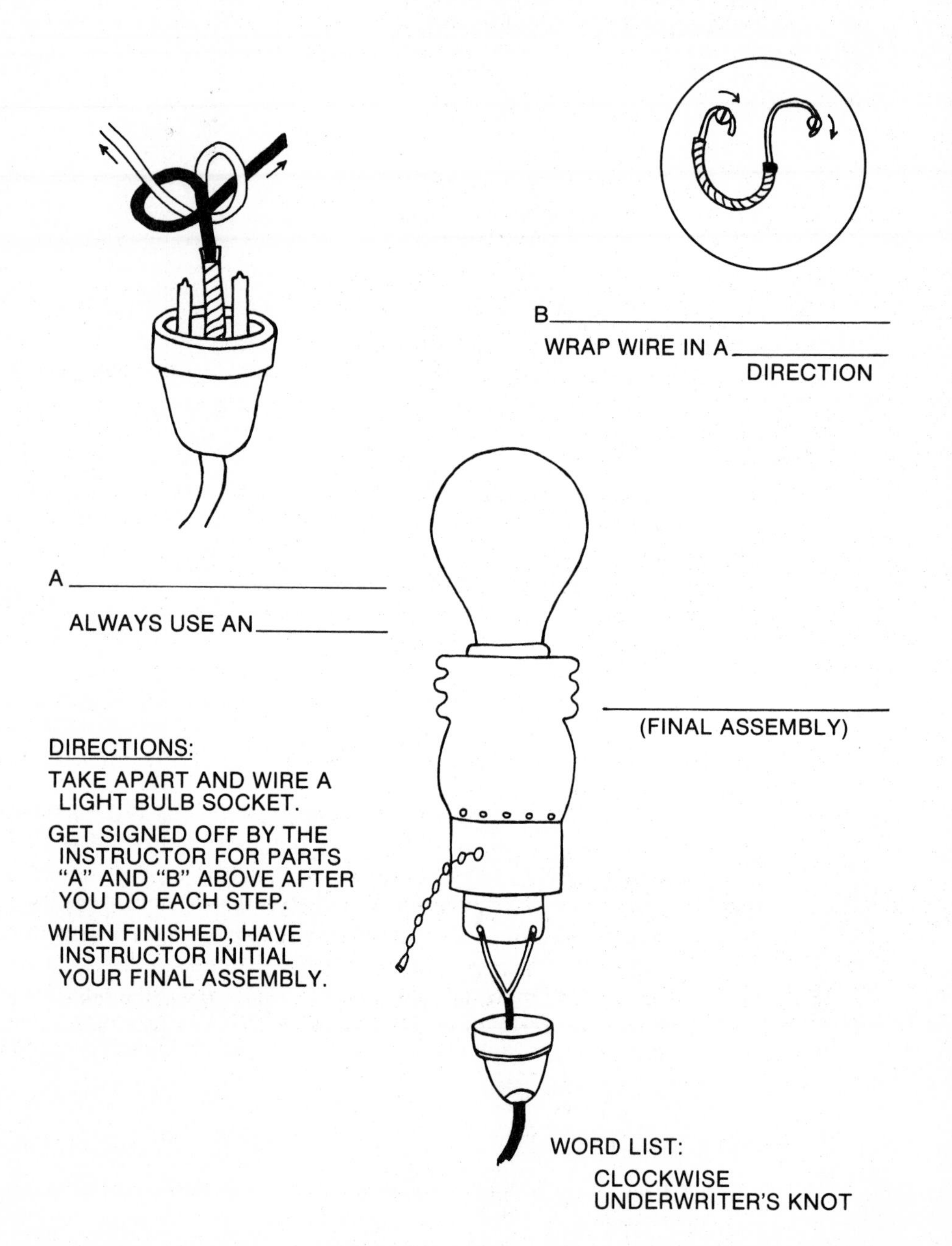
B____________________
WRAP WIRE IN A__________
DIRECTION
A____________________
ALWAYS USE AN__________
____
(FINAL ASSEMBLY)
DIRECTIONS:
TAKE APART AND WIRE A LIGHT BULB SOCKET.
GET SIGNED OFF BY THE INSTRUCTOR FOR PARTS "A" AND "B" ABOVE AFTER YOU DO EACH STEP.
WHEN FINISHED, HAVE INSTRUCTOR INITIAL YOUR FINAL ASSEMBLY.
WORD LIST:
CLOCKWISE
UNDERWRITER'S KNOT

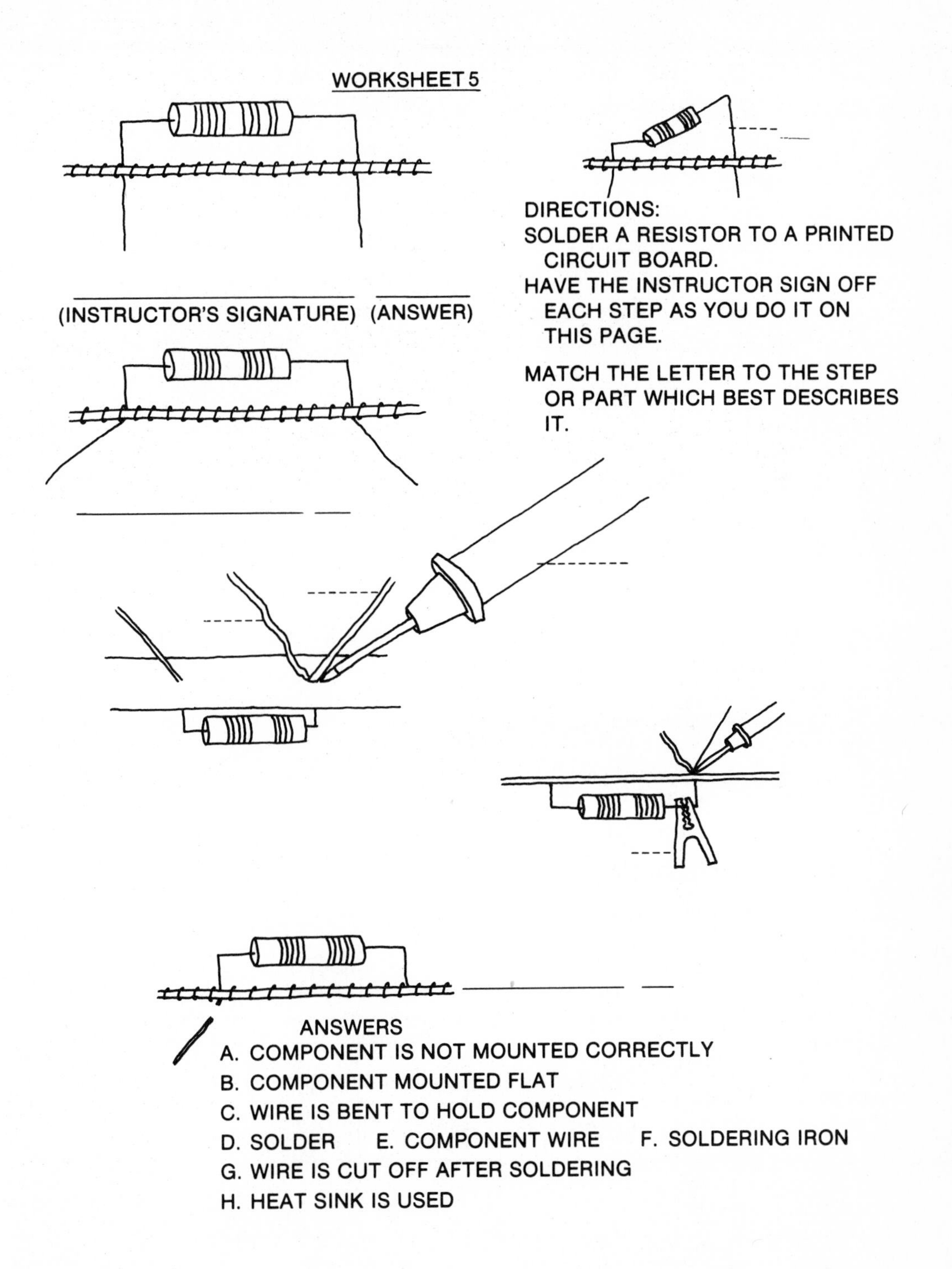

WORKSHEET 5

DIRECTIONS:
SOLDER A RESISTOR TO A PRINTED CIRCUIT BOARD.
HAVE THE INSTRUCTOR SIGN OFF EACH STEP AS YOU DO IT ON THIS PAGE.

MATCH THE LETTER TO THE STEP OR PART WHICH BEST DESCRIBES IT.

______________________ ________
(INSTRUCTOR'S SIGNATURE) (ANSWER)

______________________ ________

______________________ ________

ANSWERS
A. COMPONENT IS NOT MOUNTED CORRECTLY
B. COMPONENT MOUNTED FLAT
C. WIRE IS BENT TO HOLD COMPONENT
D. SOLDER E. COMPONENT WIRE F. SOLDERING IRON
G. WIRE IS CUT OFF AFTER SOLDERING
H. HEAT SINK IS USED

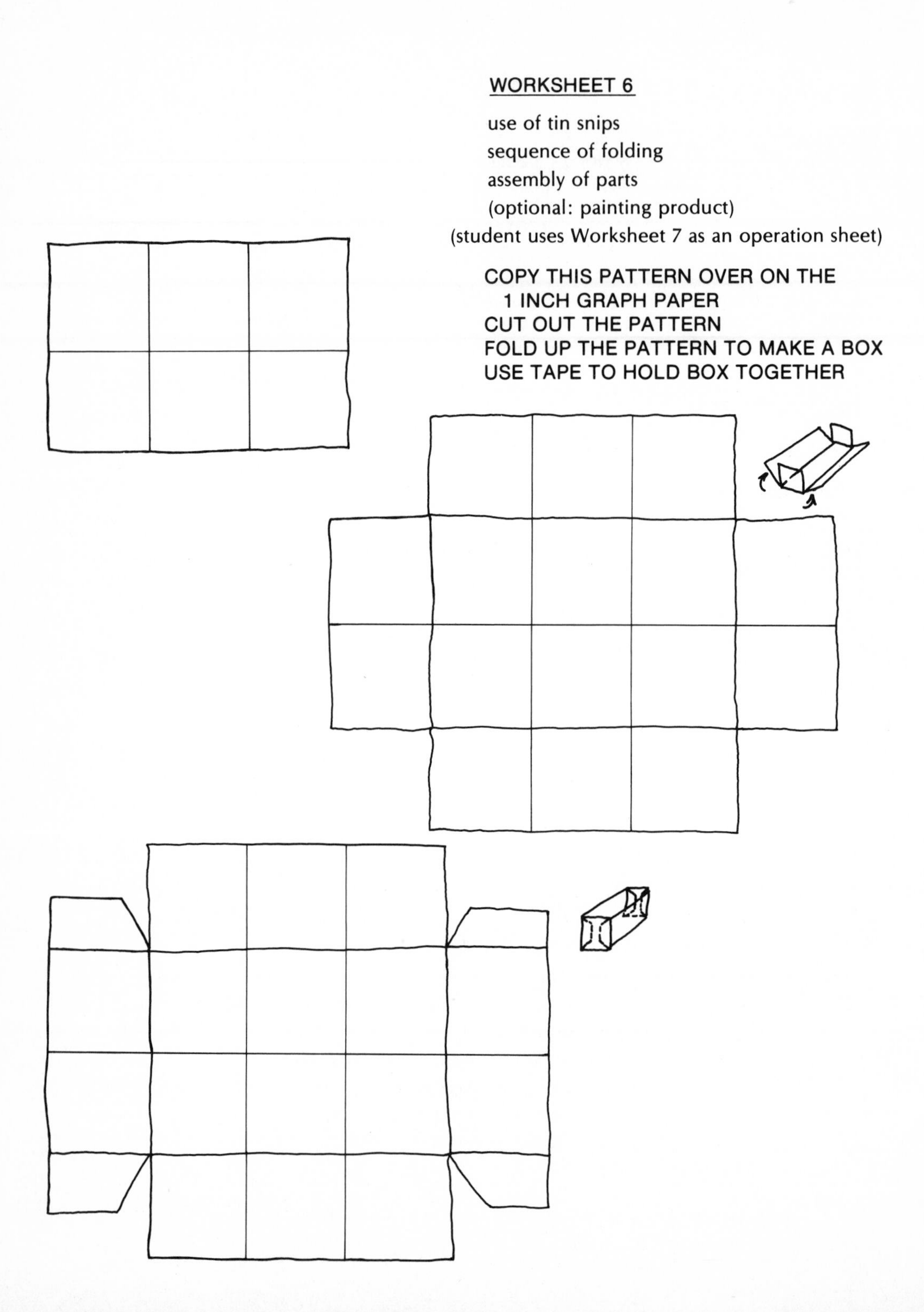

WORKSHEET 6

use of tin snips
sequence of folding
assembly of parts
(optional: painting product)
(student uses Worksheet 7 as an operation sheet)

COPY THIS PATTERN OVER ON THE
1 INCH GRAPH PAPER
CUT OUT THE PATTERN
FOLD UP THE PATTERN TO MAKE A BOX
USE TAPE TO HOLD BOX TOGETHER

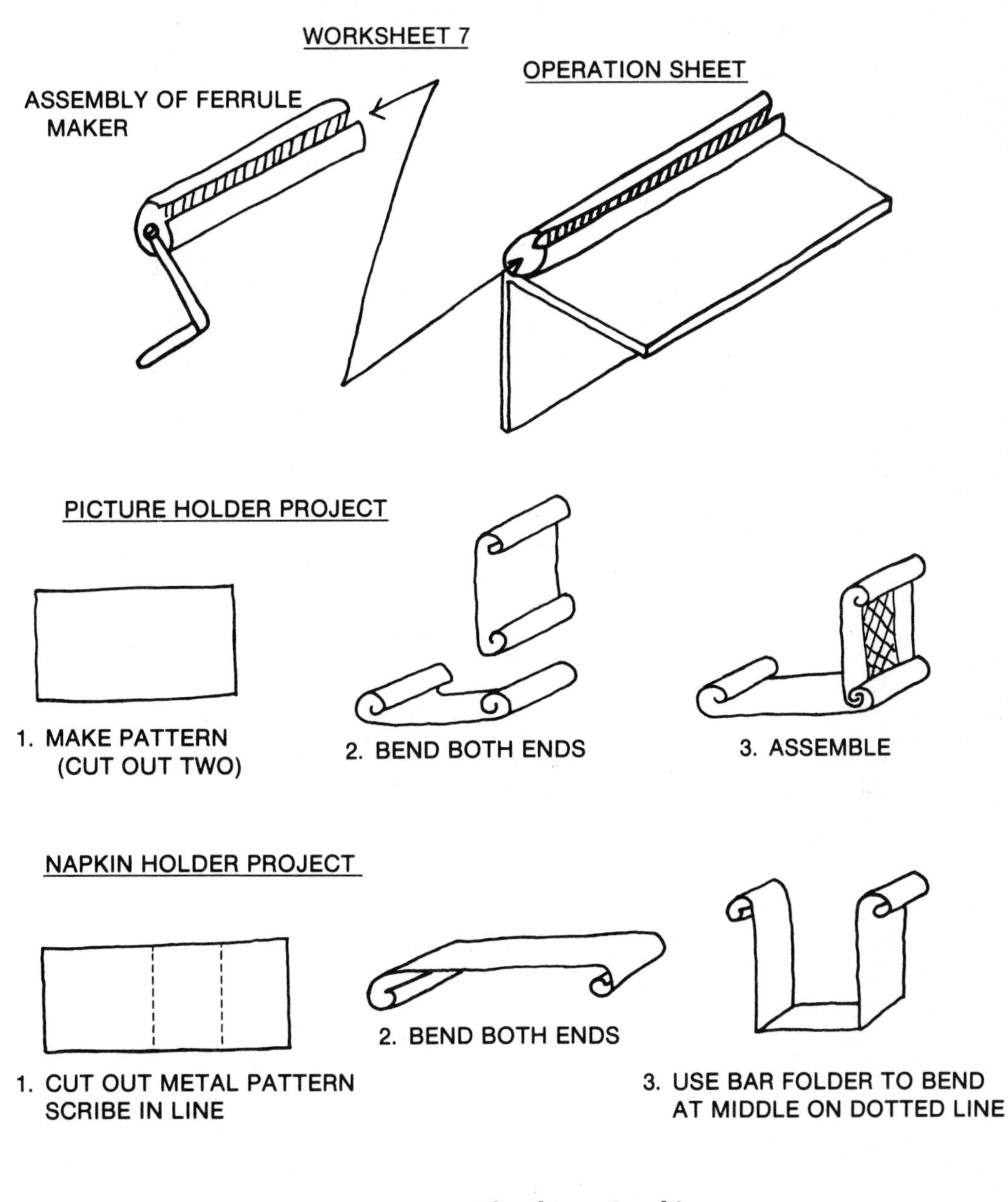
WORKSHEET 7
OPERATION SHEET
ASSEMBLY OF FERRULE MAKER
PICTURE HOLDER PROJECT
1. MAKE PATTERN (CUT OUT TWO)
2. BEND BOTH ENDS
3. ASSEMBLE
NAPKIN HOLDER PROJECT
1. CUT OUT METAL PATTERN SCRIBE IN LINE
2. BEND BOTH ENDS
3. USE BAR FOLDER TO BEND AT MIDDLE ON DOTTED LINE

Technology: Graphics

Goal	Explore photographic methods of reproduction
Product	Picture of student
Materials	Enlarger and paper processor in light-tight box (see pictures in chapter 5, "Small Group Exploratory Instructional Modules")
	Chemicals

	Photographic paper Camera and film or negatives Extension cord and outlet
Lesson	
Concepts	Reproduction of an image Photo-sensitive chemicals and paper Explore by making images using lamp Explore processes in a darkroom if available Graphics design Consider lay-out concepts, look at magazine ads Consider qualities of a good photograph Camera functions Enlarger functions
Demonstration	Loading a camera with film Use of camera Loading an enlarger with negatives Loading frame with light-sensitive paper Use of paper processor

Technology: Graphics

Goals	Explore beginning drafting tools and processes
Product	Ability to follow directions, use drafting tools
Materials	Worksheets (only two are presented here; with other worksheets the materials list would be longer) Straight edge and T-square or equivalent Pencil Drafting table or board
Lesson	
Concepts	Straight edge Parallel lines Multiviews (separate lessons) Orthographic projection (previous lessons involving gradual movement from three-dimensional representation to two-dimensional would be necessary for this lesson to have full impact)
Demonstrations	Making parallel lines Orthographic projects based on 45-degree line (prerequisites include use of multiviews incorporating front, top, and right side views) (Student completes Worksheets 8 and 9)

WORKSHEET 8

USING A T-SQUARE AND STRAIGHT EDGE
DRAW PARALLEL VERTICAL LINES
AT EACH DOT ON THE BOTTOM LINE
POINTS A TO J

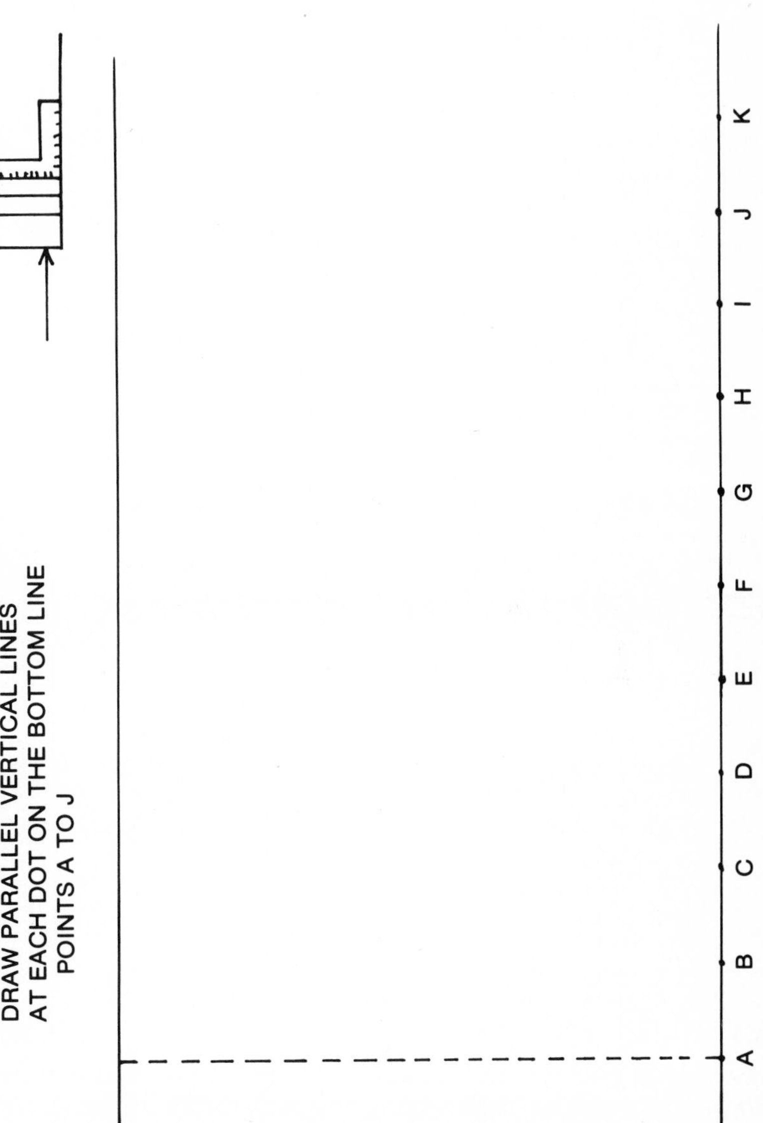

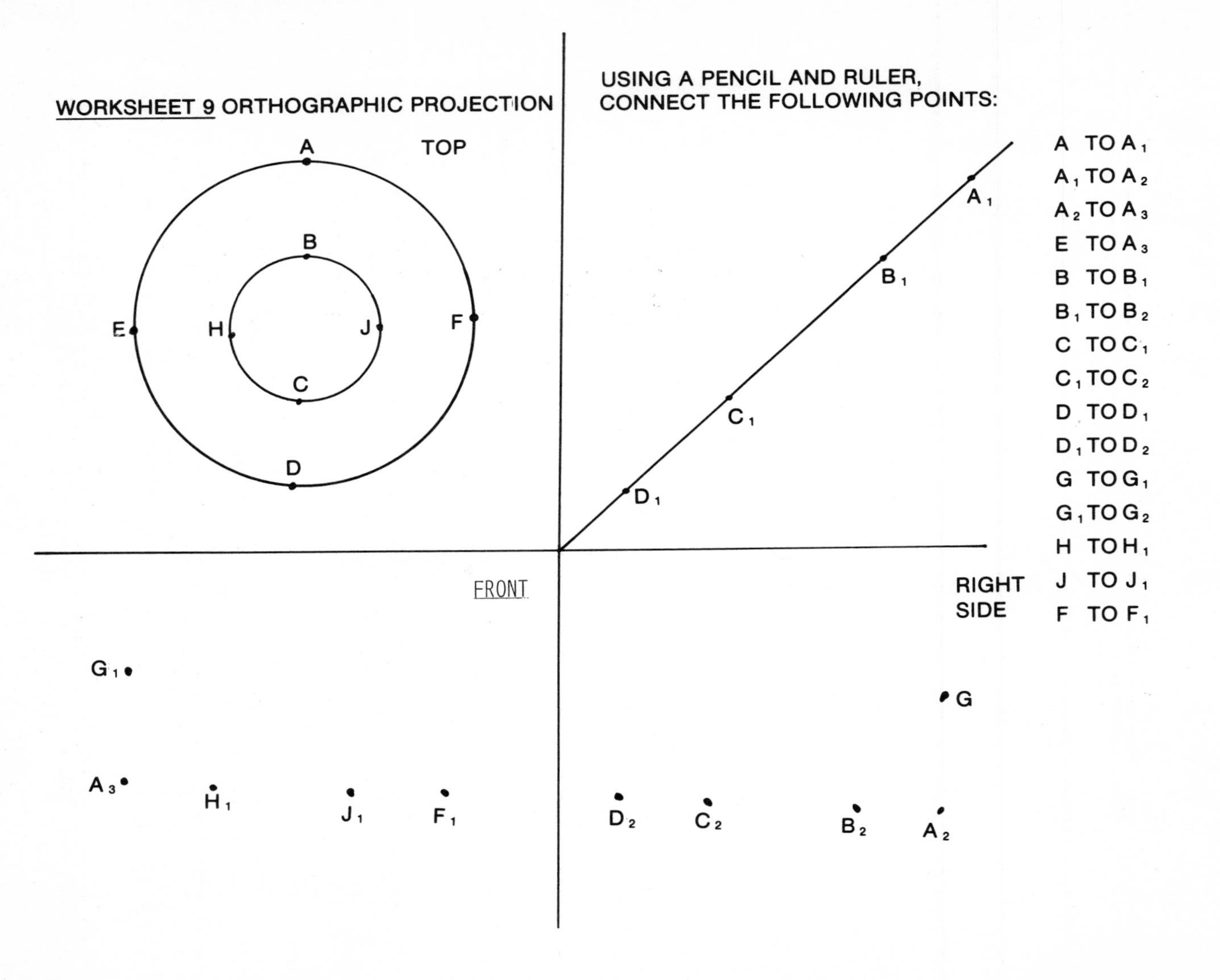
WORKSHEET 9 ORTHOGRAPHIC PROJECTION
TOP
A
B
E
H
J
F
C
D
USING A PENCIL AND RULER,
CONNECT THE FOLLOWING POINTS:
A₁
B₁
C₁
D₁
A TO A₁
A₁ TO A₂
A₂ TO A₃
E TO A₃
B TO B₁
B₁ TO B₂
C TO C₁
C₁ TO C₂
D TO D₁
D₁ TO D₂
G TO G₁
G₁ TO G₂
H TO H₁
J TO J₁
F TO F₁
FRONT
RIGHT
SIDE
G₁
G
A₃
H₁
J₁
F₁
D₂
C₂
B₂
A₂

Appendix C

RESOURCES

CURRICULUM MATERIALS ON SPECIAL EDUCATION OR SPECIAL NEEDS LEARNERS

Directories

"Annotated Bibliography of Low-Cost Vocationally Oriented Materials for Adolescent and Young Adult Mildly Handicapped and Disadvantaged Individuals." Freddie W. Litton and Richard S. Kay. *The Journal for Vocational Special Needs Education*. 2:2, 13–8, Jan. 1980.

Learning Resource Directory Westinghouse Corporation (see local libraries).

Materials Development Center (MDC) Stout Vocational Rehabilitation Institute, School of Education, University of Wisconsin—Stout, Menomonie, WI 54751.

The National Center for Research in Vocational Education Ohio State University, 1960 Kenny Road, Columbus, OH 43210 (800 -848-4815).

New Jersey Vocational-Technical Curriculums Laboratory Rutger—The State University, Building 4103, Kilmer Campus, New Brunswick, NJ 08903

National Information Center for Special Educational Materials (NICSEM) University of Southern California, University Park (RAN), 2nd Floor, Los Angeles, CA 90007 (800-421-8711).

Vocational Instruction Materials for Students with Special Needs Northwest Regional Educational Laboratory, 700 Lindsay Boulevard, 710 SW 2nd, Portland, OR 97209.

Vocational Resource Materials for Special Education R. H. Lomber, L. W. Tindal, et al. Center for Studies in Vocational-Technical Education, University of Wisconsin, Madison, WI 53706.

Instructional Programs (See Chapter 5)

Basic Skill: Singer, 80 Commerce Drive, Rochester, NY 14623 (programmed instruction with audiovisual format).

CRI: Capital Area Career Center, 611 Hagdorn Road, Mason, MI 48854 (instructional modules).

Ideal Developmental Labs, 3044 South 92nd Street, West Allis, WI 53227. Building and home repair (programmed instruction with audiovisual format).

Ken Cook: See *Technology Instructor.* Small engines and welding. 12855 W. Silver Spring Drive, P.O. Box 207, Butler, WI 53007. (programmed instruction with audiovisual format).

Project Discovery: Southwest Iowa Learning Resources Center, 401 Reed Street, Red Oak, IA 51566 (instructional modules plus hardware suitable for exploratory courses and junior high school).

Vocational Assessment and Development Program: Broadhead-Garret Co., 161 Commerce Circle, P.O. Box 15528, Sacramento, CA 95815 (curriculum materials and hardware from vocational evaluation through entry-level and cluster-level training).

ZVI Method: Wood Technology Instructional Packages, P.O. Box 3711, Portland, OR 97208.

Literature

Bulletins on Science and Technology for the Handicapped, American Association for the Advancement of Science, Washington, D.C.

Career Development for Exceptional Individuals, CEC, 1920 Association Drive, Reston, VA 22091.

Education Unlimited, Educational Resources Center, 1834 Meetinghouse Road, Boothwyn, PA.

The Exceptional Parent, 296 Boylston Street, Boston, MA 02116.

Industrial Education, Harcourt, Brace, Jovanovich, New York.

Journal of Special Needs, The Center for Vocational Personnel Preparation, Reschini House, Indiana University of Pennsylvania, Indiana, PA 15705.

Man/Society/Technology, American Industrial Arts Association, Inc., 1201 16th Street N.W., Washington, D.C. 20036.

Pathfinder, National Rehabilitation Information Center, Washington, D.C.

Life-Centered Educational Materials (commercial)

Note: Many of the materials listed under "Life-Centered Educational Materials (commercial)" contain instructional materials related to shop safety, usually about hand tools, equipment, housekeeping, and shop attitudes.

Basic Living Skills, Interpretive Education Products, Inc., 2306 Winters Drive, Kalamazoo, MI 49002.

Basic Living Skills, Survival Skills Competency Based Education catalog, Life Skill Co., 380 Maple Avenue, W., Vienna, VA 22180.

Career *Aids, Inc.* Various catalogs. 8950 Lurline Avenue, Department S8, Chatsworth, CA 91311.

Lakeshore Lifeskills. Catalog and products. 2695 E. Domingues Street, P.O. Box 6261, Carson, CA 90749.

Survival Education Catalog. Janus Book Publisher, 2501 Industrial Parkway, W. Hayward, CA 94545.

Career Education (see also chapter 5 references)

Career and Vocational Education for Small Schools; Salem, Ore: Career Education Department, no date.

Competency-Based Model for the Handicapped Adult: Living Skills and Vocational Preparation. Olympia, Wa.: Special Services Section, Washington Department of Education, n.d.

Development of a Plan for Providing Career Information for Handicapped Students. Tacoma, Wa.: Fort Steilacoom Community College, Office of Occupational Education, n.d.

Hall, Betty L. *Survival Education Oregon*. Salem, Oregon: Thomas Binford, 1977.

Kachel, Kathy. *Individual Performance Checklist: Basic Skills for the Trainable High School Student*. Ill.: The Illinois Network of Exemplary Occupational Education Programs for Handicapped and Disadvantaged, n.d. Illinois State University, Normal, Ill. 61761.

Kimeldorf, Martin. *Job Search Education*. New York: Educational Design, Inc. 1983.

Roskos, Frank. *Preparing for the World of Work*. Merrill, Wis.: F.R. Publications, 1979.

Stutrud, Carolyn. *Pre-employment Competencies*. Crestwood, Ill.: Sauk Area Career Center, Handicapped/Disadvantaged Project, n.d.

Vocational-Technical and Related Education (see also chapter 5 references)

Brown, R. N. *Development of Curriculum for a Non-Traditional Machine Tool Technology Program Accessible to the Physically Handicapped.* Hayward, Ca.: Chabot-College.

Bruewelheide, K. L. *Assisting the Physically Handicapped: An Identification and Development of Apparatus for Laboratory Shop Phase I.* Bozeman, Mont.: Montana State University, Department of Agricultural and Industrial Education, 1979.

College Guide for Students with Disabilities. Cambridge, Mass.: ABT Books, 1976.

Hohensil, T.H., and Maddy-Bernstein, C. *Resource Guide: Vocational Counselling for the Handicapped*. Blacksburg, Va.: Virginia Polytechnic Institute and State University, College of Education, 1980.

Jamison, S.I.L., ed. *Computing Careers for Deaf People*. New York: Association for Computing Machinery, Inc., 1976.

Meers, G. D., ed., *Handbook of Special Vocational Needs Education*. Gaithersburg, Md.: Aspen Systems Corp., 1980.

Phelps, L.A., and Batchelor, L.T. *Individualized Education Programs (IEP): A Handbook for Vocational Education*. Columbus, Ohio: National Center for Research in Vocational Education, 1979.

A Resource Directory of Handicapped Scientists. Washington, D.C.: American Association for Advancement of Science, n.d.

School Shop 37:8, April 1978.

Urbana Bureau of Education Research, *Preparing Vocational and Special Education Personnel to Work with Special Needs Students*. Urbana, Ill.: University of Illinois, n.d.

Vocational Development of Special Needs Individuals. Albany, Ore.; Linn-Benton Community College, n.d.

Weisgerber, Robert. *Mainstreaming the Handicapped in Vocational Education*. Ca.: Behavioral Science and Technology Group, 1977.

Wisconsin Vocational Studies Center. "Educational Programs for Persons with Special Needs." *Vocational Materials*. Madison, Wis.: University of Wisconsin—Madison, 1981. This is a catalog.

Wisconsin Vocational Studies Center. *Identifying Handicapped Students and Their Vocational Needs*. Madison, Wis.: University of Wisconsin—Madison, Wis.

Safety Instructional Materials

Kimbrell, Grady, and Vineyard, Ben S. *Entering the World of Work*, Bloomington, Ill. McKnight (Chapter 15, "General Safety in the Community" and Chapter 7, "Safety on the Job"), 1975.

Schilit, Jefferey, and Pace, Tommy J. *High Risk Employment and the Mentally Retarded Student*. CDEI, (CEC-DCD) pp. 97–103, Fall 1978. (Article on use of color codes for instruction in high-risk work environments.)

Washington State Industrial Arts Safety Guide, Seattle and Bellevue School Districts, Sept. 1976. (provides safety program guide, pictures of safe and unsafe conditions, etc.)

Films:

"Basic Practices," "Hand Tools," and "Power Tools" by Coronet Films, 369 West Erie Street, Chicago, Ill. Covers protective clothing, transporting materials, tools, housekeeping, storage, etc. See review in *Industrial Education*, 68:7, 26, Oct. 1979.

Specialized Training Projects and Rehabilitation Engineering

Electronic Industries Foundation, Project with Industry, 2001 Eye Street, N.W., Washington, DC 20006, with offices in California and Massachusetts. The project attempts to train people to work in electronics industries.

National Rehabilitation Information Center, 4407 Eighth Street, N.E., The Catholic University of America, Washington, DC 20017. Funded by Rehabilitation Services Administration to disseminate information to the rehabilitation community regarding research, funding, bibliographies, locating factual information, and statistics.

Texas Rehabilitation Commission, 118 E. Riverside Drive, Austin, TX 78704. Engineers attempt to overcome difficulties encountered at the worksite and at home by locating existing adapted technology or creating new ones.

Rehabilitation Engineering Centers:

Case Western Reserve University, 2219 Adelbert Road, Cleveland, OH 44106.

National Institute for Rehabilitation Engineering, 97 Decker Road, Butler, NJ 07405.

New York University, Medical Rehabilitation T and T Center, 400 East 14th Street, New York, NY 10016.

Northwestern University, 435 East Superior Street, Room 1441, Chicago, IL 60611.

Smith-Kettlewell Institute of Visual Sciences, 2232 Western Street, San Francisco, CA 94115.

University of Iowa, Orthopedics Department, Dill Children's Hospital, Iowa City, IA 52242.

University of Tennessee, Department of Orthopedic Surgery, 1248 LaPaloma Street, Memphis, TN 38114

Washington Department of Services for the Blind. 3411 South Alaska St., P.O. Box 18379, Seattle, WA 98118. Has the latest equipment for accommodating blind people at the worksite and has excellent information for job site modification.

Tools, Equipment and Machinery Adapted for the Vocational Education and Employment of Handicapped People. Compiled by Jon Gugerty, Arona Faye Roshal, Mary D.J. Tradewell, and Linda Anthony. Catalog number SNE301, Wisconsin Vocational Studies Center, University of Wisconsin—Madison, Publications, 265 Educational Sciences Building, Madison, WI 53706. A catalog of materials useful to teachers, consumers, and employers.

ORGANIZATIONS (SEE CHAPTERS 3 AND 4)

Special-Technology Education

Division for Career Development, Council for Exceptional Children, 1920 Association Drive, Reston, VA

National Association of Vocational Educators Special Needs Personnel (NAVESNP), American Vocational Association, 2020 North 14th Street, Arlington, VA 22201

Special Needs Division, American Vocational Association, 2020 N. 14th St., Arlington, Virginia 22201

Technology Education

American Association for the Advancement of Science, 1515 Massachusetts Avenue, N.W., Washington, DC 20005

American Industrial Arts Association, 1914 Association Drive, Reston, VA 22091

American Vocational Association, 2020 N 14th, Arlington, VA 22201

Special Education Advocacy

Federal Agencies

Architectural and Transportation Barriers Compliance Board, 330 C Street, S.W., Washington, DC 20202

Crippled Children's Services, Office of Maternal and Child Health, Bureau of Community Health Services, HEW, 5600 Fishers Lane, Room 7-15 Parklawn Building, Rockville, MD 20852

Developmental Disabilities Office, Office of Human Development, HEW, Room 3070 Switzer Building, Washignton, DC 20201

Library of Congress, Division for the Blind and Physically Handicapped, Taylor Street Annex, 1291 Taylor Street, Washington, DC 20542

Office of Handicapped Individuals, Humphrey Building, 200 Independence Avenue, S.W., Washington, DC 20201

President's Committee on Employment of the Handicapped, Washington, DC 20202

Rehabilitation Services Administration, Office of Human Development, HEW, Room 4324 Switzer Building, Washington, DC 20201

White House Conference on Handicapped Individuals, 1832 M Street, N.W., Washington, DC 20036.

Private National Organizations

Alexander Graham Bell Association for the Deaf, Washington, DC

Allergy Foundation of America, New York, NY

American Alliance for Health, Physical Education, Recreation and Dance, Washington, DC

American Association for Children with Learning Disabilities, Pittsburgh, PA

American Association for Gifted Children, New York, NY

American Association for the Education of the Severely/Profoundly Handicapped, Seattle, WA

American Association of University Affiliated Programs for the Developmentally Disabled, Washington, DC

American Association on Mental Deficiency, Washington, DC

American Coalition of Citizens with Disabilities, Washington, DC

American Diabetes Association, New York, NY

American Occupational Therapy Association, Rockville, MD

American Physical Therapy Association, Washington, DC

American Speech and Hearing Association, Washington, DC

American Printing House for the Blind, Louisville, KY

Association for the Help of Retarded Children, New York, NY

Association for the Visually Handicapped, Louisville, KY

Arthritis Foundation, New York, NY

Boy Scouts of America, Scouting for the Handicapped Division, North Brunswick, NJ

Closer Look, National Information Center for the Handicapped, P.O. Box 1492, Washington, DC

Council for Exceptional Children, 1920 Association Drive, Reston, VA

Cystic Fibrosis Foundation, Atlanta, GA

The Dysautonomia Foundation, Inc., New York, NY

Epilepsy Foundation of America, Washington, DC

Friedreich's Ataxia Group in America, Oakland, CA

International Association of Parents of the Deaf, Silver Spring, MD

John Tracy Clinic, Los Angeles, CA (deafness/hearing impairments, deaf-blind)

League for Emotionally Disturbed Children, New York, NY

Leukemia Society of America, New York, NY

Little People of America, Owatonna, MN

Mental Retardation Association of America, Salt Lake City, UT

Muscular Dystrophy Association, Inc., New York, NY

National Amputation Foundation, Whitestone, NY

National Association for Retarded Citizens, Arlington, TX

National Association for Visually Handicapped, New York, NY

National Association of Physically Handicapped, London, OH

National Ataxia Foundation, Minneapolis, MN

National Center for a Barrier-Free Environment, Washington, DC

National Center for Law and the Deaf, Washington, DC

National Center for Law and the Handicapped, South Bend, IN

National Easter Seal Society for Crippled Children and Adults, Chicago, IL

National Epilepsy League, Chicago, IL

National Federation of the Blind, Des Moines, IA

National Foundation/March of Dimes, White Plains, NY

National Hearing Aid Society, Livona, MI

National Hemophilia Foundation, New York, NY

National Multiple Sclerosis Society, New York, NY

National Paraplegia Foundation, Chicago, IL

National Society for Autistic Children, Huntington, WV

National Tuberous Sclerosis Association, Laguna Beach, CA

Orton Society (dyslexia), Towson, MD

Osteogenesis Imperfecta Foundation, Burlington, NC

Prader-Willi Association, Long Lake, MI

Sex Information and Education Council of the US (SIECUS), Hempstead, NY

Society for the Rehabilitation of the Facially Disfigured, New York, NY

Spina Bifida Association of America, Madison, WI

United Cerebral Palsy Associations, Inc., Los Angeles, CA

United Ostomy Association, Inc., Los Angeles, CA

Western Law Center for the Handicapped, Los Angeles, CA

VIP COMMUNICATOR

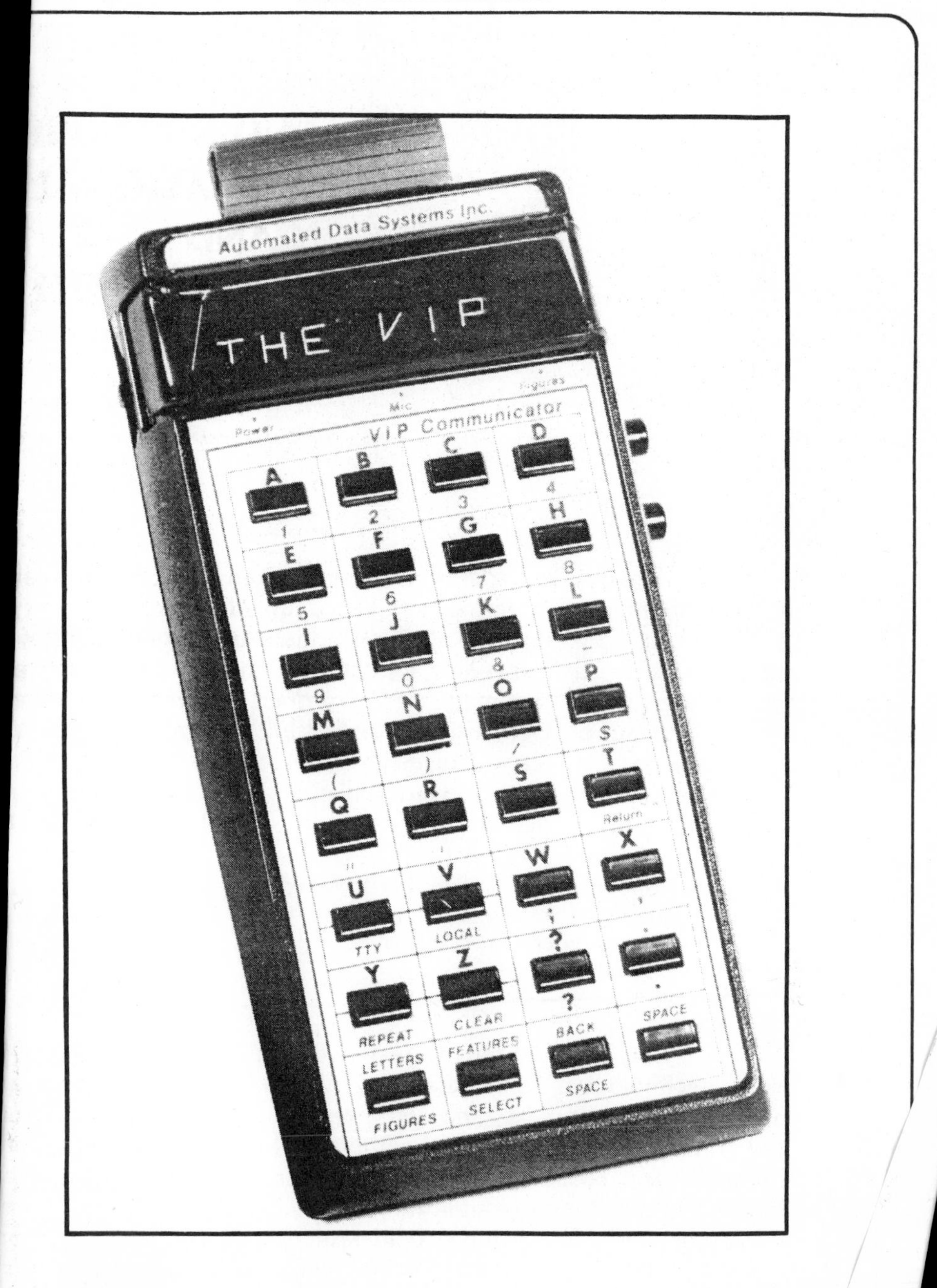

otograph courtesy of Sue
lke, ADS, Inc.

Appendix L

TOOLS, EQUIPMENT, AI ADAPTED FOR THE V EDUCATION EMPLOYMENT OF HA PEOPLE

The following materials are samples tak lection of material of the same title compil Faye Roshal, Mary D. J. Tradewell, and Linda related to arts, gardening, sewing, typing, cor cation. The material was made under contract cation (Bureau of Occupational and Adult Vocational Studies Center, University of Wiscoı subcontract with the Wisconsin Indianhead Adult Education District, Shell Lake. Inquiries sł

Wisconsin Vocational Studies Center Publicatio
265 Educational Sciences Building
University of Wisconsin—Madison
Madison, WI 53706
Phone: (608) 263-4357

Materials were selected for this text to demonst tional technologies. Some relate to communica as prosthetics or guides, and some relate to m It was the intent of the original authors of these primarily, and specific information for purchas terested in acquiring or duplicating the example disclaimers regarding patents, field testing, end stated by the original authors but not included h are included as ideas, not as solutions. Several e ety of problems (e.g., tool adjusting blocks can limited measuring abilities).

Ph
En

VIP COMMUNICATOR

DEVELOPER

Automated Data Systems, Inc.
P.O. Box 4062
Madison, WI 53711
(608) 273-0707 (Voice or TTY)

CONTACT PERSON

Same as Developer

WHERE IT IS USED

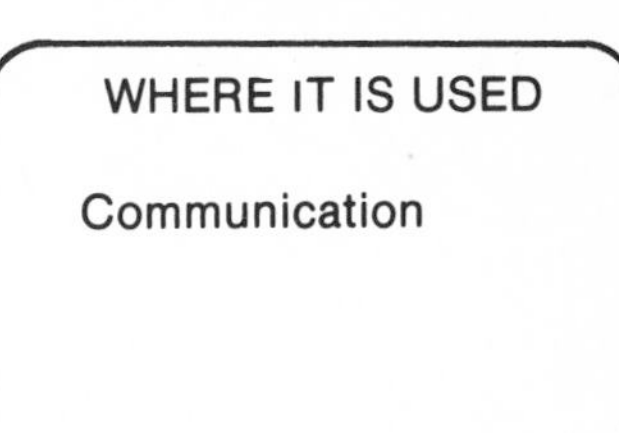

Communication

PROBLEM(S) IT OVERCOMES

difficulty experienced by deaf persons in communicating by phone

FIELD TESTED

information not available

REGULATORY APPROVAL

information not available

WARRANTY PROVIDED

yes

FOR SALE

Automated Data Systems, Inc.
P.O. Box 4062
Madison, WI 53711

VIP: $179.00 + shipping
Talking Pocket: $79

HOW IT WORKS

Features - The VIP weighs 9 oz. It has an automatic carriage-return and linefeed for use with printing TTYs, a backspace key, a light to tell you when there is a dial tone, ringing, or busy signal on the telephone, a clear key for clearing the display, a "memory" that lets you backspace up to 45 characters to review a message. Because question marks and periods are used so often, the VIP lets you type them without needing to type the figures key and then the letters key (VIP's computer does it for you). Emergency messages of up to 45 characters may be entered in VIP's memory. By pressing the features and then the repeat keys, the message will be sent over and over. This lets you attend to the emergency while the person receiving your call can get help. Option - The Talking Pocket. This device is used to "talk" with other people face to face. The Talking Pocket may be worn on your shirt pocket, blouse, or lapel. When words are typed on the VIP, they appear on the Talking Pocket as well. The person you are "talking" to simply reads the words from your pocket. This is very useful when talking to hearing persons since they often do not understand sign language. The VIP comes complete with rechargeable batteries, AC adapter, telephone connectors, carrying case, and instruction booklet. The optional Talking Pocket may be ordered separately.

(Information based on company literature.)

AM-COM I (TTY)

Photograph courtesy of American Communication Corporation

AM-COM I (TTY)

DEVELOPER

American Communication Corporation
180 Roberts Street
East Hartford, CT 06108
(203) 289-3491
(Voice or TTY)

CONTACT PERSON

Same as Developer

WHERE IT IS USED

Communication

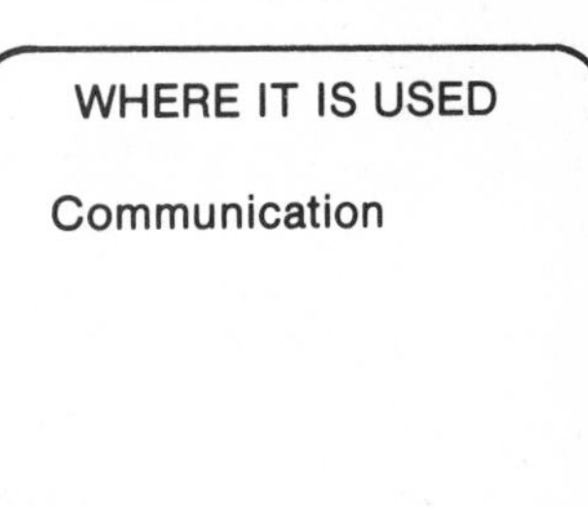

PROBLEM(S) IT OVERCOMES

inability of someone who is deaf to communicate by phone

FIELD TESTED

information not available

REGULATORY APPROVAL

information not available

WARRANTY PROVIDED

yes

FOR SALE

American Communication Corporation
180 Roberts Street
East Hartford, CT 06108

AM-COM 1	$495
Shipping	$ 10
Emergency message feature	$ 65
Carrying case	$ 25

HOW IT WORKS

AM-COM 1 is battery operated and will function without being plugged into an electrical outlet. Now you can call from a pay phone, at work, or even while visiting. AM-COM 1 is compatible with most automatic phone answering equipment. In addition to AM-COM 1's 900 character built-in memory storage system, unlimited memory is possible with the use of a supplementary tape recorder. In the event of an emergency and help is needed, dial the emergency number and depress the Emergency Key and 1 key for police, 2 key for fire, or 3 key for ambulance. AM-COM 1 will automatically send a TTY message for help repeatedly with your name and address while you handle the emergency.

(Information based on company literature.)

SENSORY QUILL

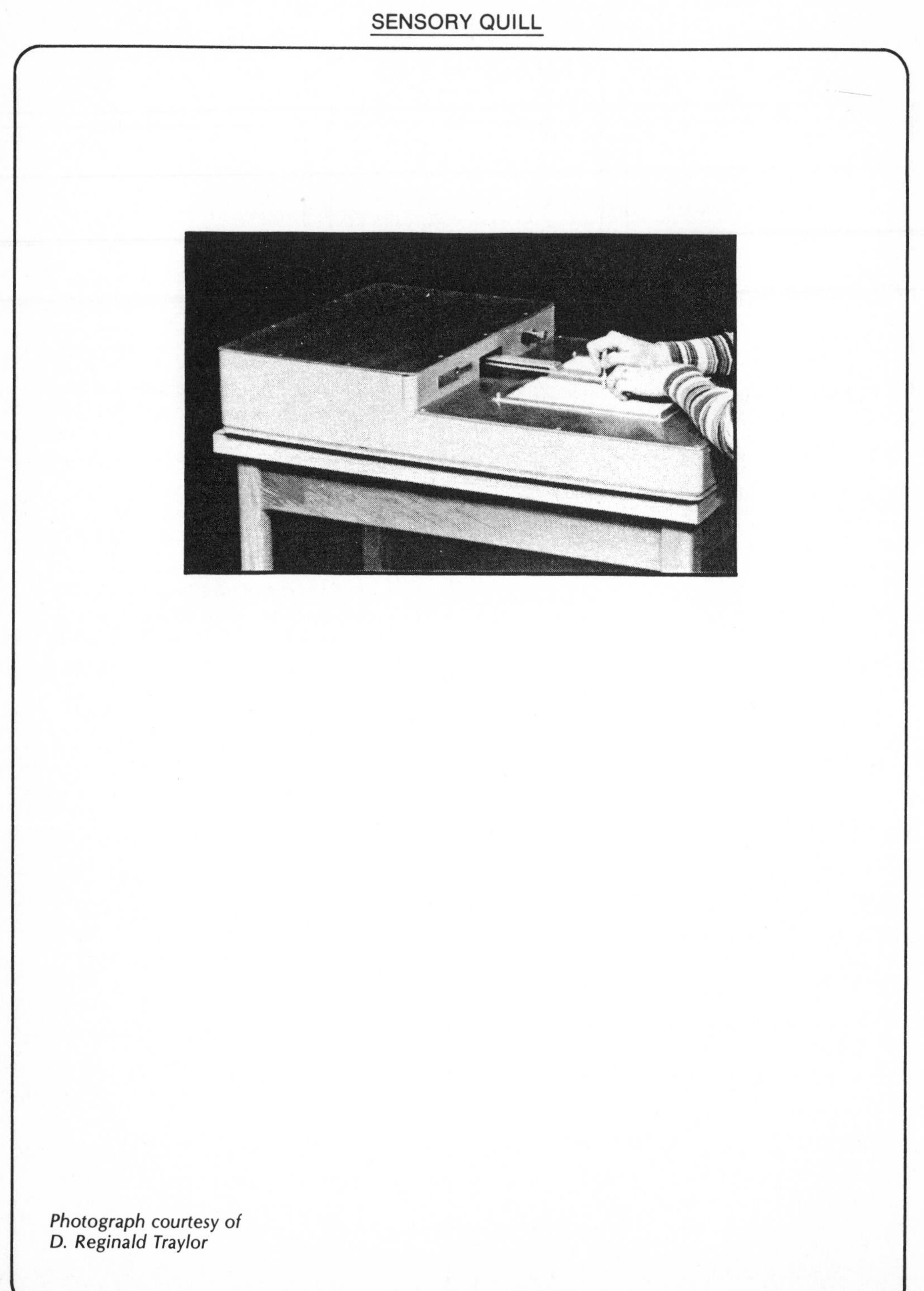

Photograph courtesy of D. Reginald Traylor

SENSORY QUILL

DEVELOPER

Traylor Enterprises, Inc.
830 N.E. Loop 410
Suite 505
San Antonio, TX 78209
(512) 828-0203

CONTACT PERSON

D. Reginald Traylor
or M.L. Jones
Traylor Enterprises, Inc.
830 N.E. Loop 410
Suite 505
San Antonio, TX 78209
(512) 828-0203

WHERE IT IS USED

Arts
Construction
Drawing
Information Processing
Writing

PROBLEM(S) IT OVERCOMES

inability of a visually impaired person to see drawings, graphs, plans, maps, etc.

FIELD TESTED

yes

REGULATORY APPROVAL

not applicable

WARRANTY PROVIDED

yes

FOR SALE

Traylor Enterprises, Inc.
830 N.E. Loop 410
Suite 505
San Antonio, TX 78209
Institutional Model - $795
Personal Model - $425

HOW IT WORKS

The user moves a stylus across the writing and drawing area. Wherever the stylus touches the paper or plas-uic, a raised image occurs. Script, graphics, drawings, and maps can be produced quickly and easily in raised form, with no inversion necessary for tactual inspection. Whatever a sighted person sees in the nature of drawings, graphs, plans, maps, etc. can now be tactually "seen" by the visually handicapped. Importantly, the visually handicapped can generate their own raised line drawings, better communicating with the nonhandicapped.

(Information based on Traylor Enterprises, Inc. literature.)

STANLEY CALIPER RULE

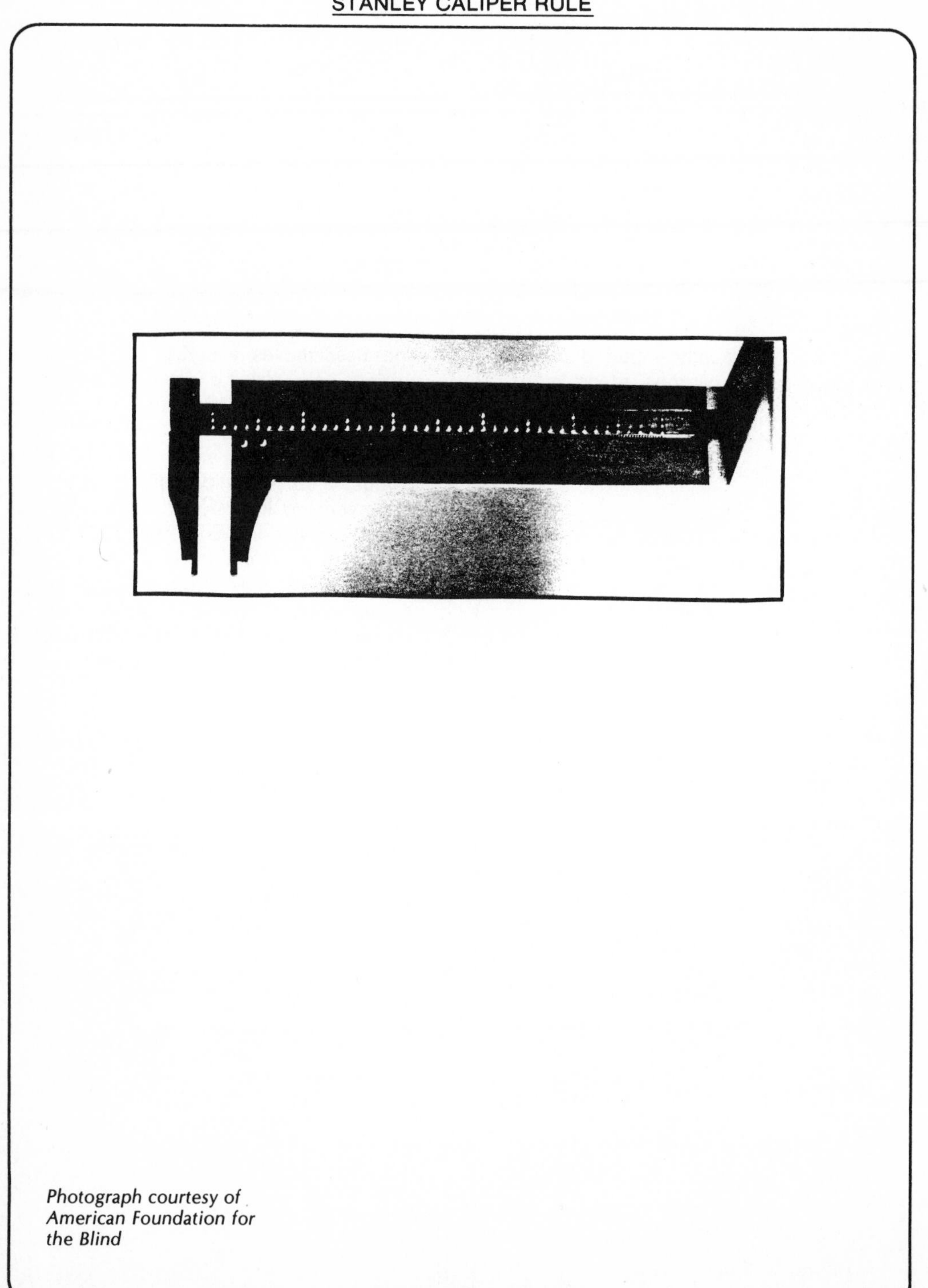

Photograph courtesy of American Foundation for the Blind

STANLEY CALIPER RULE

DEVELOPER

CONTACT PERSON

Alex H. Townsend
American Foundation for the Blind
15 West 16th Street
New York, NY 10011
(212) 620-2169

WHERE IT IS USED

Benchwork
Construction

PROBLEM(S) IT OVERCOMES

visual impairment

FIELD TESTED

yes
by panels of blind and visually impaired persons

REGULATORY APPROVAL

information not available

WARRANTY PROVIDED

information not available

FOR SALE

American Foundation for the Blind
15 West 16th Street
New York, NY 10011

$19

HOW IT WORKS

The sliding jaw has a raised scale graduated in 1/8″ marks indicated by a single raised dot. For quick reading, the ½″ mark has double raised dots and the 1″ mark has three raised dots for inside and outside measurement. The stationary jaw has double raised dots on the face to indicate inside and outside measurements.

(Information based on American Foundation for the Blind literature.)

AUD-A-METER

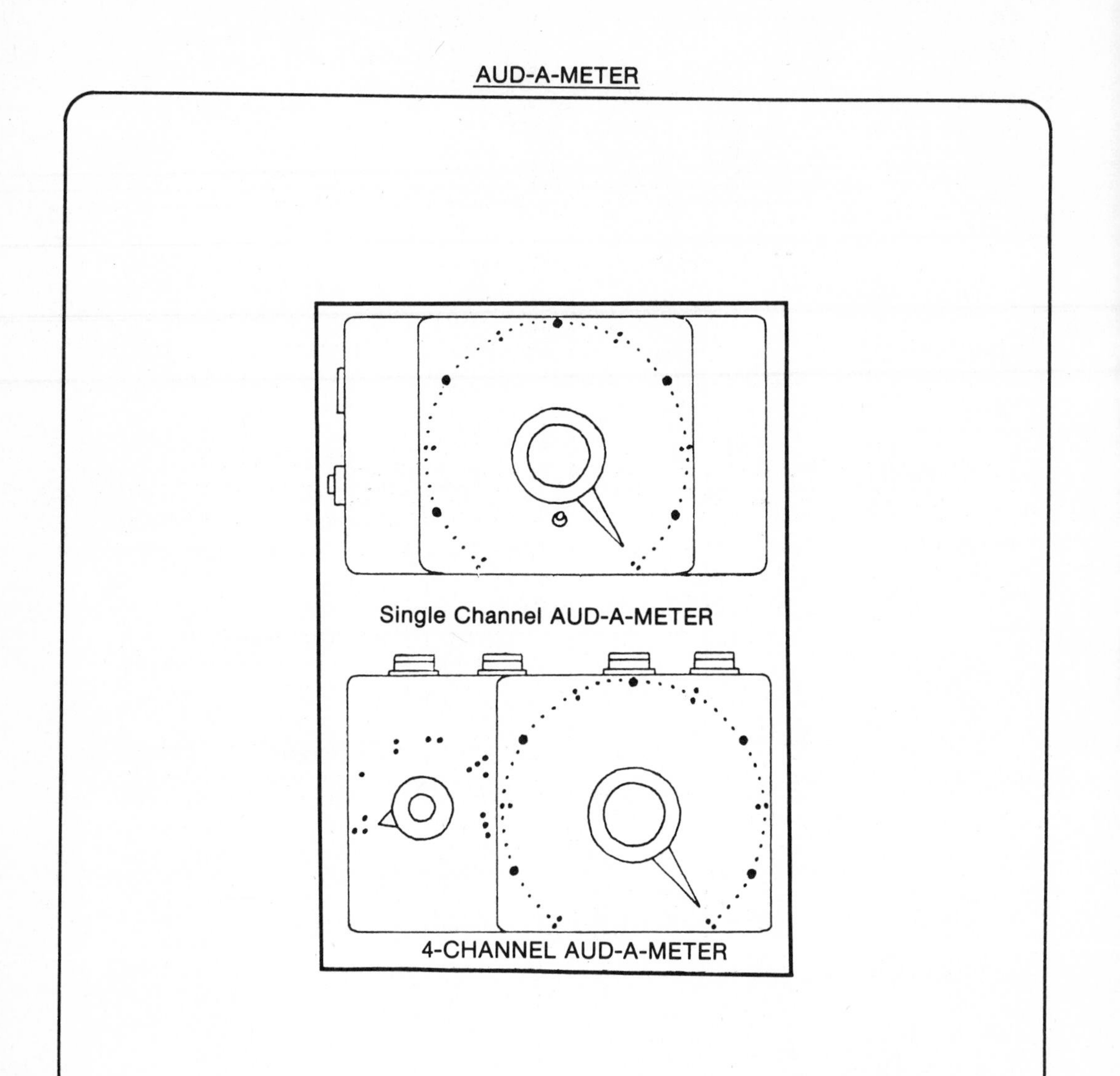

Illustration courtesy of Science for the Blind Products

AUD-A-METER

DEVELOPER

CONTACT PERSON

Tom Benham
Science for the Blind Products
Box 385
Wayne, PA 19087
(215) 687-3731

WHERE IT IS USED

Meter reading

PROBLEM(S) IT OVERCOMES

visual impairment

FIELD TESTED

information not available

REGULATORY APPROVAL

information not available

WARRANTY PROVIDED

information not available

FOR SALE

Science for the Blind Products
Box 385
Wayne, PA 19087

Single chann 1 AUD-A-METER	$ 90
4-Channel AUD-A-METER	$215

HOW IT WORKS

The Single Channel Aud-A-Meter may be connected across the terminals of any electrically driven visual meter movement in order to allow reading to be taken by correlating an auditory signal with a braille scale. Sensitivity of approximately 150 millivolts allows the Aud-A-Meter to be used satisfactorily with almost any visual meter movement, DC application only. The 4-channel Aud-A-Meter can be connected to four different meters at once and is particularly applicable to broadcast use, where several meters must be monitored.

(Information based on Science for the Blind literature.)

GRINDER MODIFICATION I

Photograph courtesy of A.R. Colby

GRINDER MODIFICATION I

DEVELOPER

Pratt & Whitney Aircraft Group
Manufacturing Division
400 Main Street
East Hartford, CT 06108

CONTACT PERSON

A.R. Colby, Manager
EED Program
Pratt & Whitney Aircraft Group
Manufacturing Division
400 Main Street
East Hartford, CT 06108

WHERE IT IS USED

Manufacturing

PROBLEM(S) IT OVERCOMES

inability of someone who is legally blind to adjust a typical grinder machine quickly and accurately

FIELD TESTED

yes

used by one employee

REGULATORY APPROVAL

information not available

WARRANTY PROVIDED

information not available

FOR SALE

For information about obtaining a closed circuit TV camera and viewing screen, contact:

Visualtek
1610 26th Street
Santa Monica, CA 90404

HOW IT WORKS

The grinder operator pictured opposite is legally blind (20/200 vision or less when wearing glasses). He can operate 5-6 pieces of milling and grinding machinery. To accommodate him, Pratt & Whitney obtained a few special aids. The grinder operator uses a conventional 50X microscope to check pieces for chips or other flaws. He uses a couple of double magnifying lenses to check the "Last Word" indicastor of his grinder. He also uses a closed circuit television camera and viewing screen made by Visualtek to check his work. The Visualtek camera can be seen in the upper left corner of the picture and is positioned directly on his work. The viewing screen can be seen just behind his right shoulder.

(Information provided A.R. Colby.)

MEASURING DEVICE FOR WOODWORKING: DADOING

Photograph courtesy of Russ Gage

MEASURING DEVICE FOR WOODWORKING: DADOING

DEVELOPER

information not available

CONTACT PERSON

Russ Gage
4820 South 20th Street
Milwaukee, WI 53221
(414) 281-0076

WHERE IT IS USED

Carpentry
Sheet Metal Working
Woodworking

PROBLEM(S) IT OVERCOMES

difficulty experienced by a blind person when making precise cuts in wood or other materials

HOW IT WORKS

Each piece in this set is 6″ in length. They are used to line up a saw or a router. The *thickness* is the unit of measurement. The smallest is 1/8″ thick; the next is 2/8″ ot 1/4″ in width, and so on up to 1″. (Eight pieces of metal are fastened together to make the 1″ piece.) The grooves running across each piece allow a blind user to make sure he or she is using the desired size. (Four grooves equals 4/8″ or 1/2″, etc.) If you must dado a board to a depth of, say, 3/8″ you can set the 3/8″ unit nut to the table saw blade and raise or lower the blade to the desired height.

(Information provided by Russ Gage.)

FIELD TESTED

yes

used by one blind person

REGULATORY APPROVAL

not applicable

WARRANTY PROVIDED

not applicable

FOR SALE

no

TOOL ADJUSTING BLOCK, LATHE OPERATION

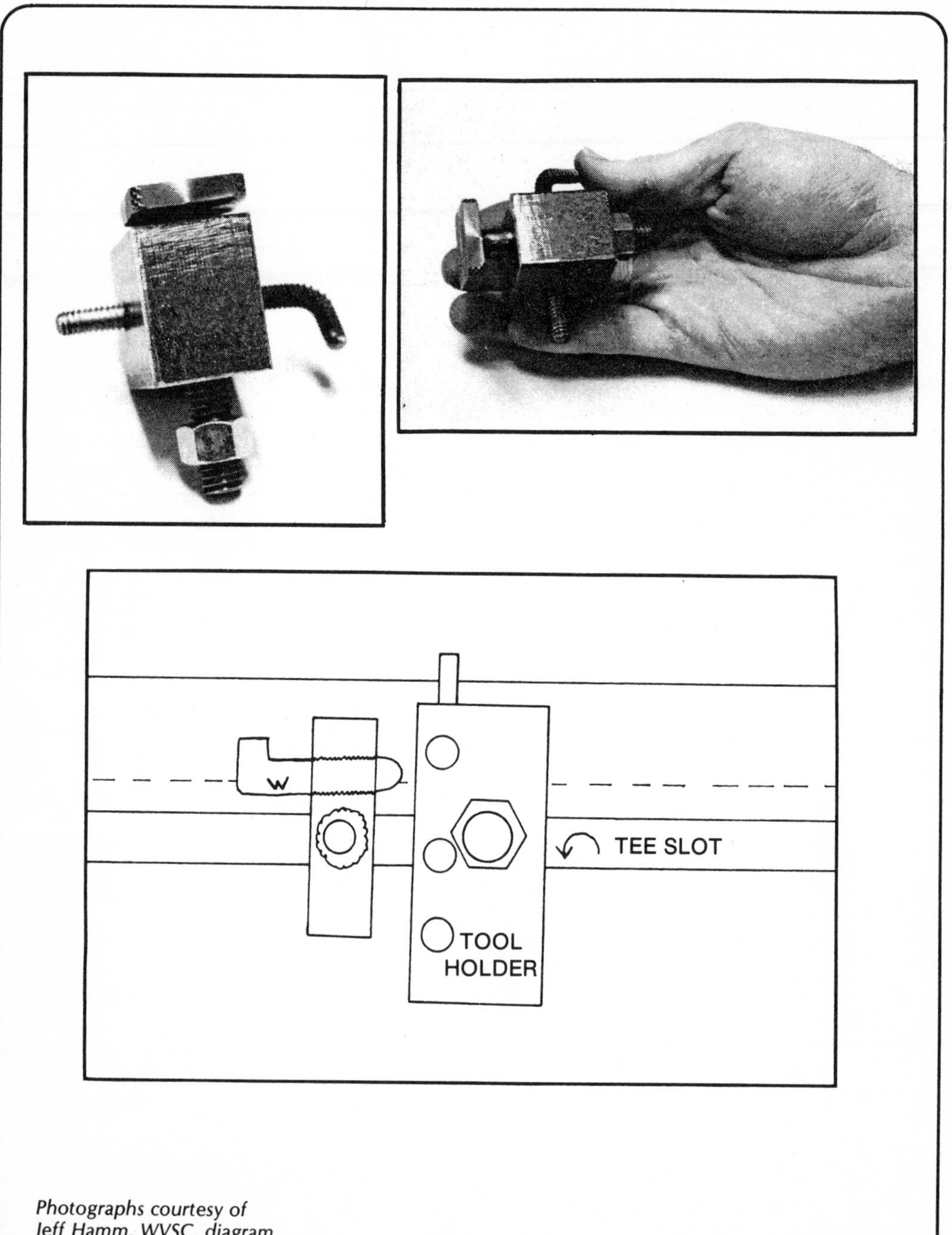

Photographs courtesy of Jeff Hamm, WVSC, diagram redrawn by Terri Bleck, WVSC

TOOL ADJUSTING BLOCK, LATHE OPERATION

DEVELOPER

Jerome Golner
Golner Precision Products
354 Cottonwood Avenue
Hartland, WI 53029

CONTACT PERSON

Same as Developer

WHERE IT IS USED

Machine Trades: Lathe Operation

PROBLEM(S) IT OVERCOMES

moving a lathe's tool holder precisely when one cannot see

FIELD TESTED

yes

used in one factory by one blind employee

REGULATORY APPROVAL

information not available

WARRANTY PROVIDED

no

FOR SALE

no

HOW IT WORKS

The tool adjusting block can be clamped at the side of a tool holder and 1/4-20 screw can be advanced or retracted to give accurate tool holder movement. For example, to move the tool holder .050 to the right, clamp the tool adjusting block next to the tool holder, turn the 1/4-20 screw until it touches the holder. Loosen the tool holder, advance the 1/4-20 screw one turn, and retighten tool holder.

(Information based on personal communication with Jerome Golner.)

A SPOT WELDER HOLDING FIXTURE

Photograph courtesy of Albert E. Swarts

A SPOT WELDER HOLDING FIXTURE

DEVELOPER

Albert E. Swarts, P.E.
Richard L. Biddy, Dir.
Vocational Industrial Center - (Houston)
Donald R. Smith, Ph.D.
Texas A & M University - (College Station)

CONTACT PERSON

Albert E. Swarts, P.E.
Vocational Industrial Ctr.
Institute for Rehabilitation and Research
2809 Main Street
Houston, TX 77002
(713) 797-1440

WHERE IT IS USED

Assembly
Benchwork

PROBLEM(S) IT OVERCOMES

dexterity limitations

FIELD TESTED

information not available

REGULATORY APPROVAL

not applicable

WARRANTY PROVIDED

not applicable

FOR SALE

May be made at a minimal cost.

HOW IT WORKS

The fixture is designed for spot welding brass connectors to both ends of a Nichrome ribbon, requiring accurate placement of the welding point. Holding fixtures, to insure proper length, are supplied but do not effect good placement of the weld. To remedy this, a frame was built around the welding tongs with a channel to center the fixture along the line of the welder tip. A stop is used to center the weld point on the other axis. This locates the fixture on 3 sides. To reduce handling and dexterity requirements, the stop is on a slide and a second one is on the other side. This permits the operator to slide the fixture through the channel to the other end instead of lifting it out and turning it around to weld the second end.

(Information based on *Examples of Jig and Fixture Design as Applied to the Severely Disabled Functioning in a Sheltered Workshop*, Biddy, Smith, and Swarts.)

MALLET, FOOT OPERATED

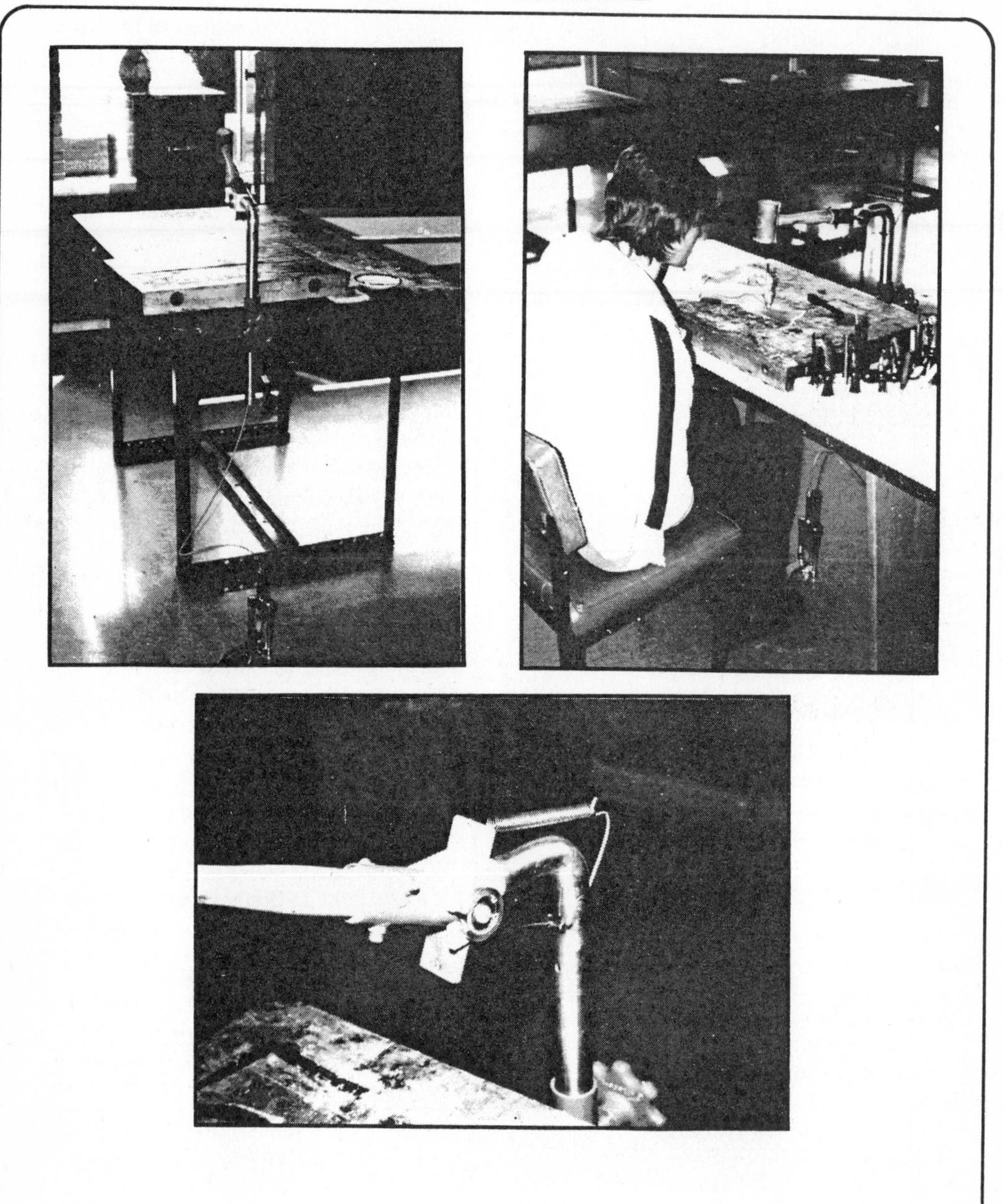

Photographs courtesy of David F. Law, Jr.

MALLET, FOOT OPERATED

DEVELOPER

David F. Law, Jr.
Woodrow Wilson Rehab. Ctr.
Rehab. Engineering Dept.
Fishersville, VA 22939
(703) 885-9724

CONTACT PERSON

Same as Developer

WHERE IT IS USED

Benchwork

PROBLEM(S) IT OVERCOMES

the need for two hands to use leather punches and stamps effectively

HOW IT WORKS

For use by hemiplegic individuals; utilizes foot depression of a pedal to swing mallet or hammer.

(Information provided by David F. Law, Jr.)

FIELD TESTED

yes

Woodrow Wilson Rehabilitation Center and Individual's Leather Shop

REGULATORY APPROVAL

information not available

WARRANTY PROVIDED

no

FOR SALE

no

PREHENSILE HAND

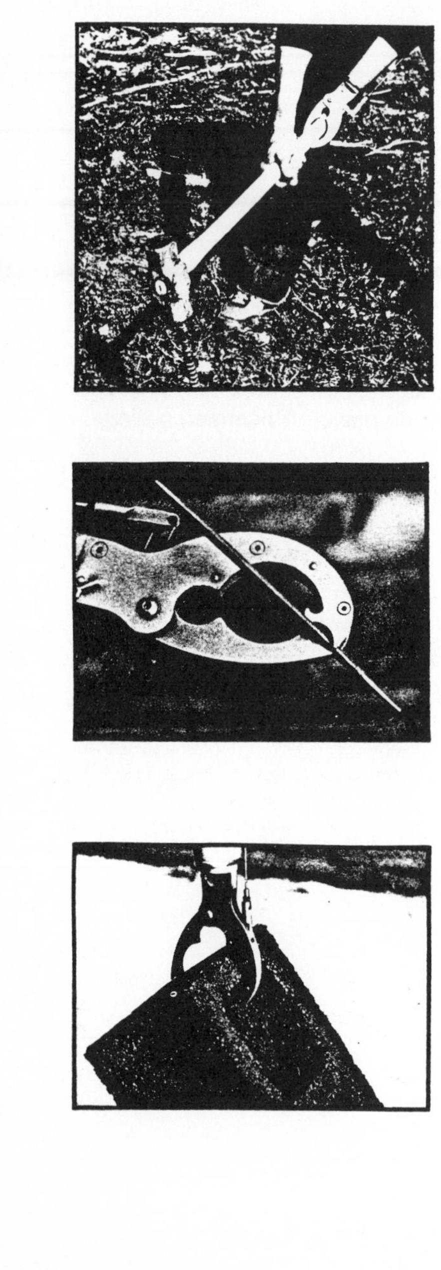

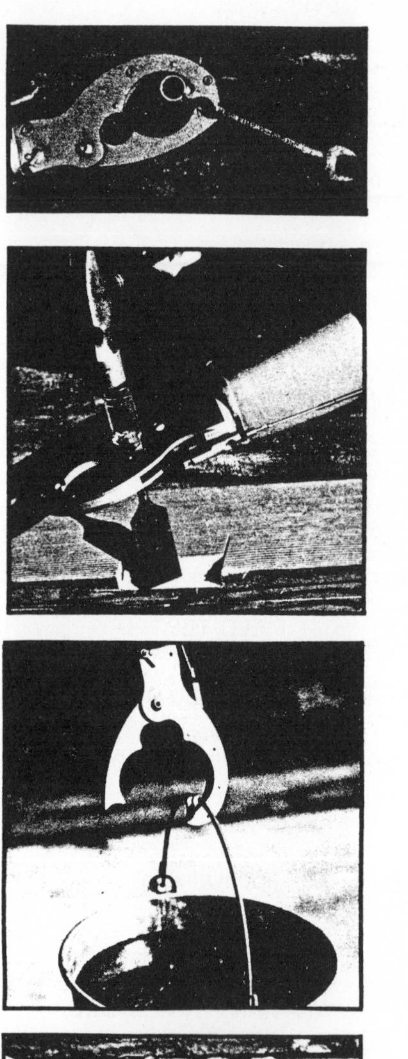

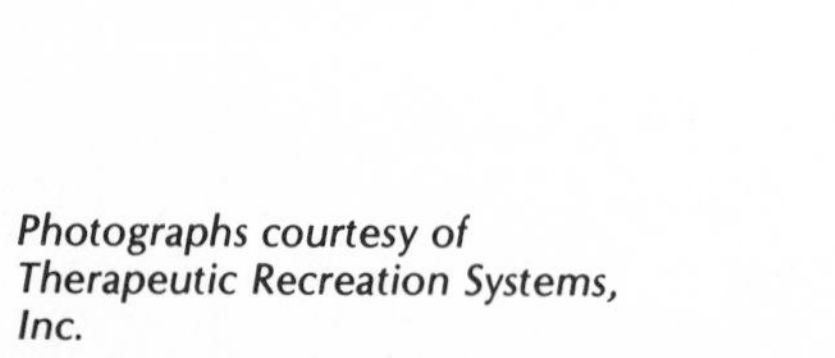

Photographs courtesy of Therapeutic Recreation Systems, Inc.

PREHENSILE HAND

DEVELOPER

Robert Radocy
1280 - 28th, Suite 4
Boulder, CO 80303
(303) 444-4720

CONTACT PERSON

Same as Developer

WHERE IT IS USED

Agriculture
Construction
Machine Trades
and others

PROBLEM(S) IT OVERCOMES

no upper limb; inability to use tools without adaptors; lack of capacity for vigorous, strenuous upper extremity activity using standard hook

FIELD TESTED

yes

REGULATORY APPROVAL

FDA

WARRANTY PROVIDED

yes

FOR SALE

Therapeutic Recreation Systems, Inc.
1280 - 28th, Suite 4
Boulder, CO 80303

or contact local prosthetist

HOW IT WORKS

Muscle powered, this device is a complete hand replacement (not applicable for persons with a partial hand amputation), which is like the functional action of a human thumb and forefingers. The Hand's thumb operates with a positive voluntary action, enabling the user to control the amount of force exerted to tighten the cable that activates the thumb. It is suggested that the locking pawls, which are manually engaged, be used for specialized activities only. By consciously gripping objects the user makes use of biofeedback that is lost when locking device is activated. The photographs depict activities accomplished without the locking device.

(Information based on Therapeutic Recreation Systems, Inc. literature.)

POSSUM™ INPUT CONTROLS

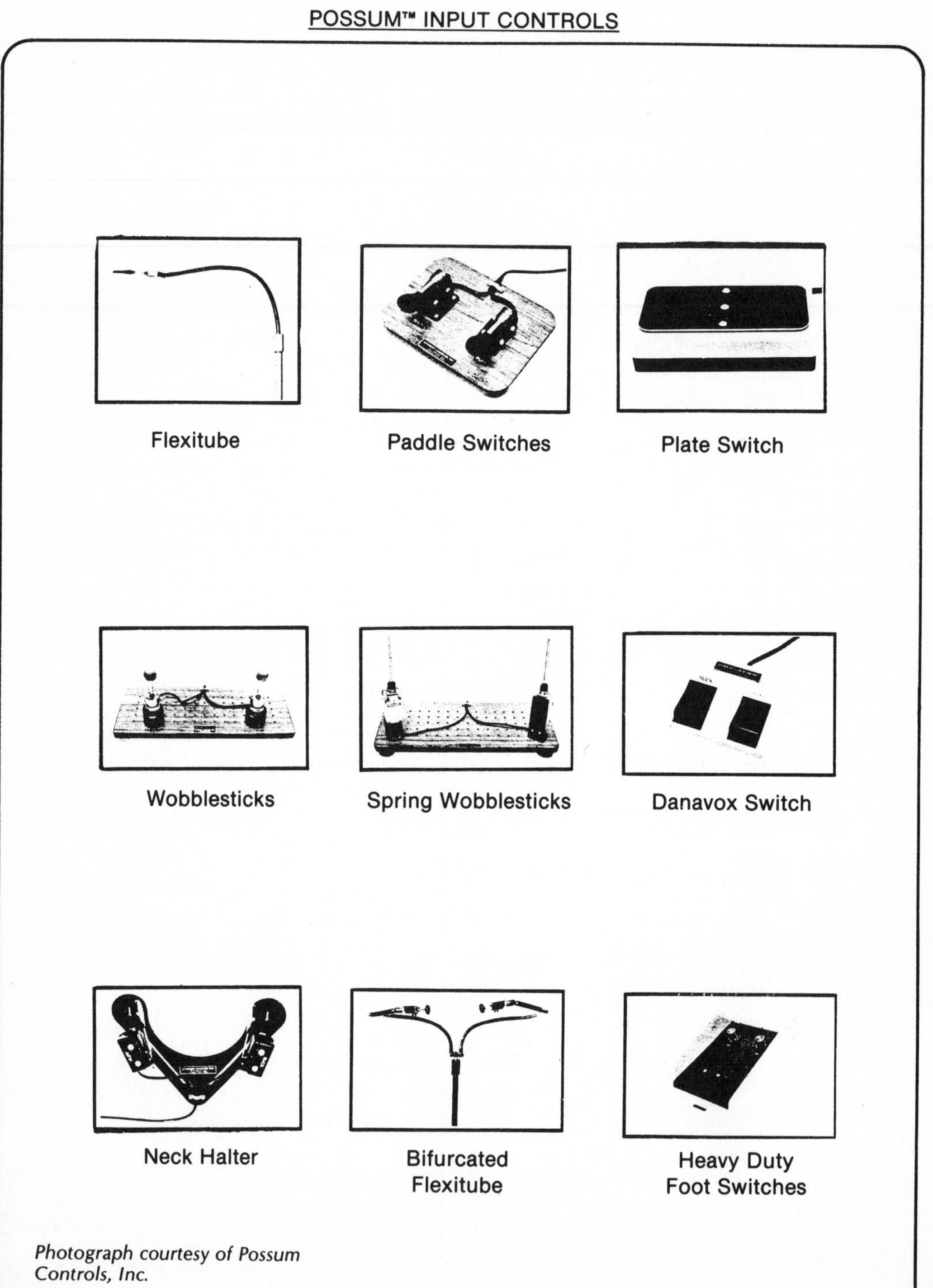

Photograph courtesy of Possum Controls, Inc.

POSSUM™ INPUT CONTROLS

DEVELOPER

Possum Controls, Inc.
111 Fairacres Industrial Estate
Windsor Berkshire
England

CONTACT PERSON

Ann F. Gurr
Possum, Inc.
P.O. Box 451
Midwood Station
Brooklyn, NY 11230
(212) 243-1658

WHERE IT IS USED

Operate Equipment

PROBLEM(S) IT OVERCOMES

controls operated by person's residual movements

FIELD TESTED

yes

Britain

REGULATORY APPROVAL

FDA

WARRANTY PROVIDED

yes

FOR SALE

Possum, Inc.
P.O. Box 451
Midwood Station
Brooklyn, NY 11230

HOW IT WORKS

Flexitube on microphone stand - free standing, operated through pressure and suction equivalent to blowing bubbles or sucking a straw, can be mounted on bed or wheelchair.
Paddle Switches - light action microswitch with paddle lever mounted on a board or unmounted for special positioning.
Plate Switch - large contact area activated through light pressure to the surface, which is divided in two, each half operating as a single switch. Single version also available.
Wobblesticks - operated by knocking them in any direction, useful for gross random movements. *Spring Wobblesticks* have springs along sticks' length to prevent injury.
Danavox Switch - used with hand or foot, needs harder pressure than other inputs.
NeckHalter - one or two light action microswitches mounted for chin operation, microswitch position individually adjustable.
Bifurcated Flexitube with microswitches on microphone stand - free standing, operated with side to side head movement, light action microswitches, single ones also available.
Heavy Duty Foot Switches - sturdy switches mounted on an angled bracket on a board for gross movement.

(Information based on Possum Controls, Inc. literature)

"MAINSTREAM" ELEVATING WHEELCHAIR

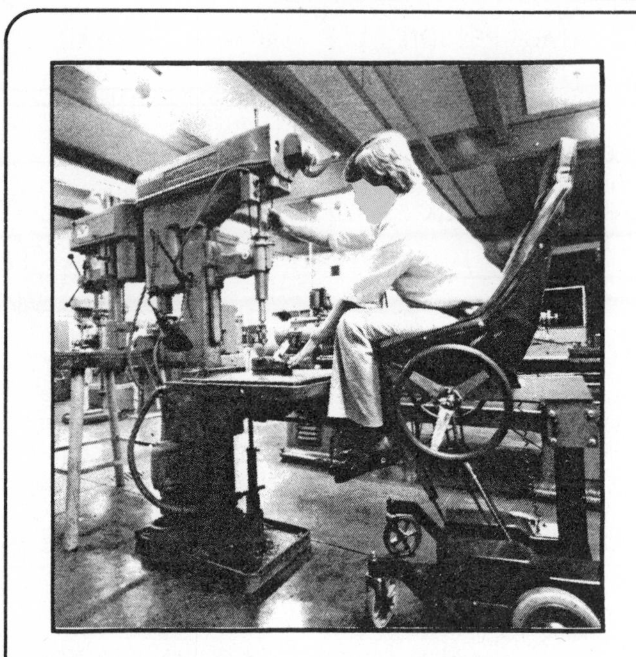

Photographs courtesy of Summit Services

"MAINSTREAM" ELEVATING WHEELCHAIR

DEVELOPER

Summit Services
535 Division Street
Campbell, CA 95008
(408) 378-1251

CONTACT PERSON

William Redmond
Glenn Brown
Summit Services
535 Division Street
Campbell, CA 95008
(408) 378-1251

WHERE IT IS USED

Manufacturing
Technical

PROBLEM(S) IT OVERCOMES

inaccessibility of many educational and work environments.

FIELD TESTED

yes

REGULATORY APPROVAL

FDA

WARRANTY PROVIDED

yes

FOR SALE

Summit Services
535 Division Street
Campbell, CA 95008

approximately $3,000

HOW IT WORKS

The following information was provided by Robert N. Brown, Chairman, Division of Technology and Engineering, Chabot College: By using this device, you adapt the person to *any* environment. At Chabot College it would have cost $15,000.00 to just adapt our machine shop, not to mention our chemistry, biology, physics, and photo labs. However, with a few of these chairs on campus we adapt each wheelchair user to any environment. The chair is comfortable, safe, easy rolling, and electrically controlled up and down. The purpose of the device is to allow persons to train on normal equipment and in normal labs so that they can easily go out into industry or business and assume high-paying jobs. The Mainstream Elevating Wheelchair is manually operated with handwheels (same manner as conventional wheelchairs), but the seat, handwheels, and footrest electronically elevate 18 inches above conventional wheelchair height. This design feature permits the occupant to be "fully mobile" at any height they wish or need to be to perform functions a standing person would perform.

This wheelchair is extremely stable at all heights and its maneuverability is excellent in small areas because it is no wider than conventional wheelchairs; in the full-up position it is 6 inches shorter because the footrest tucks under the seat as it rises.

OFFSET PRESS

RETRACTABLE HANDLE

LEVER CONTROL

Photograph courtesy of Gestetner Corporation

OFFSET PRESS MODEL NO. 319

DEVELOPER

Gestetner Corporation
Gestetner Park
Yonkers, NY 10703
(914) 968-6666

CONTACT PERSON

Joseph Schachner
or Robert Wallace
Gestetner Corporation
Gestetner Park
Yonkers, NY 10703
(914) 968-6666

WHERE IT IS USED

Communication
Duplication
Printing

PROBLEM(S) IT OVERCOMES

limited hand control

FIELD TESTED

yes

Camp Jawonio, Rockland County Center for Physically Handicapped, Inc.-New York City, NY

REGULATORY APPROVAL

UL; Canadian Standards Association

WARRANTY PROVIDED

yes

FOR SALE

Gestetner Corporation
Gestetner Park
Yonkers, NY 10703

319 Offset Duplicator	$7,050
319 Cabinet	$ 295

HOW IT WORKS

This Offset Duplicator has controls that have been designed for easy operation. Some of the special features are as follows:

Levers instead of full-grasp controls, which are easier to use.

A fold-away crank handle makes the large control wheel of the cylinder mechanism easier to manipulate.

Twist-type knobs are equipped with special extensions that permit them to be more easily turned. The covers and lids are equipped with large plastic hooks for easy handling.

Manipulation of the cylinder wheel will allow the plate to attach itself to the cylinder and will also eject itself after the run is completed. Even the printing blanket will be automatically washed and the power shut off at the completion of all necessary operations.

A "read-out" panel, which shows the progress of the duplicating process with illuminated symbols.

Other printing equipment and supplies are available from this company.

(Information based on product brochures.)

VOXCOM[R] SLOW-SPEED CARD READER AND CASSETTE RECORDER

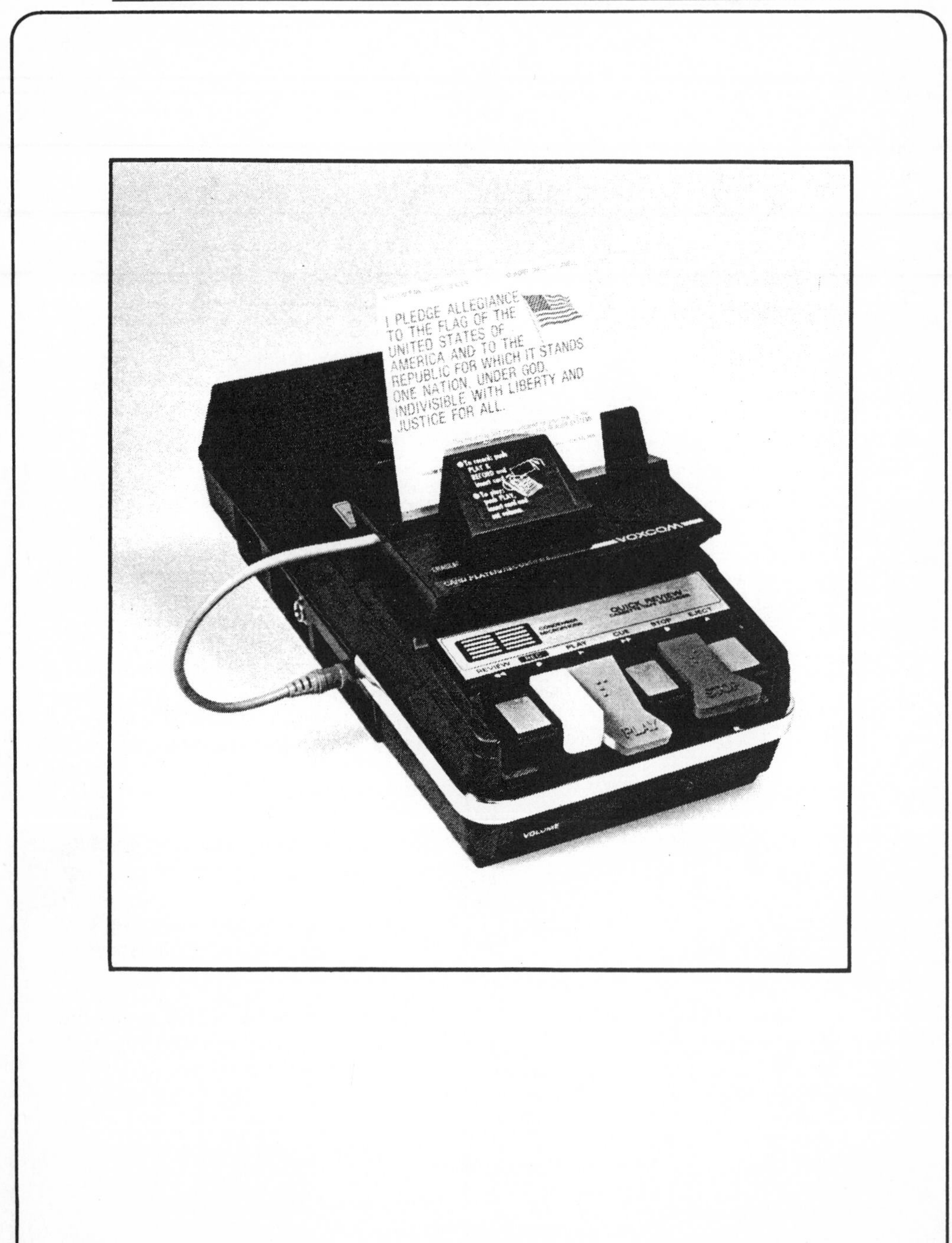

Photograph courtesy of Voxcom

VOXCOM[R] SLOW-SPEED CARD READER AND CASSETTE RECORDER

DEVELOPER

VOXCOM
Division of Tapecon, Inc.
100 Clover Green
Peachtree City, GA 30269
(404) 487-7575

CONTACT PERSON

Bonnie Pray
VOXCOM
100 Clover Green
Peachtree City, GA 30269
(404) 487-7575

WHERE IT IS USED

Reading
Teaching

PROBLEM(S) IT OVERCOMES

difficulty in reading; difficulty retaining information

FIELD TESTED

yes

in schools around the world

REGULATORY APPROVAL

no

WARRANTY PROVIDED

yes

FOR SALE

VOXCOM
Division of Tapecon, Inc.
100 Clover Green
Peachtree City, GA 30269

or call for location of nearest dealer

HOW IT WORKS

Information can be simultaneously seen, read, and heard with this unit. An adhesive-backed magnetic tape is applied to the back of any piece of paper, (e.g., the back of a photo). Information is recorded onto the magnetic tape and then played using the slow-speed card-reader adapter. By removing the adapter, the unit can be used as an ordinary cassette recorder. It can be used to teach math concepts, give machine operating instructions, and for other similar applications.
One of the VOXCOM models has color-coded extender keys with braille and tactile markings for record, play, and stop.

(Information based on VOXCOM catalog, 1980.)